ADVANCED TOPICS
IN SCIENCE AND TECHNOLOGY IN CHINA

ADVANCED TOPICS
IN SCIENCE AND TECHNOLOGY IN CHINA

Zhejiang University is one of the leading universities in China. In Advanced Topics in Science and Technology in China, Zhejiang University Press and Springer jointly publish monographs by Chinese scholars and professors, as well as invited authors and editors from abroad who are outstanding experts and scholars in their fields. This series will be of interest to researchers, lecturers, and graduate students alike.

Advanced Topics in Science and Technology in China aims to present the latest and most cutting-edge theories, techniques, and methodologies in various research areas in China. It covers all disciplines in the fields of natural science and technology, including but not limited to, computer science, materials science, life sciences, engineering, environmental sciences, mathematics, and physics.

Junbai Li

Nanostructured Biomaterials

With 122 figures, mostly in color

ZHEJIANG UNIVERSITY PRESS
浙江大学出版社

Springer

Editor
Prof. Junbai Li
Key Lab of the Colloid and Interface Sciences
International Joint Lab with the German Max
Planck Institute of Colloids & Interfaces
Institute of Chemistry
Chinese Academy of Sciences
Beijing 100190, China
E-mail: jbli@iccas.ac.cn

ISSN 1995-6819 e-ISSN 1995-6827
Advanced Topics in Science and Technology in China

ISBN 978-7-308-06601-3
Zhejiang University Press, Hangzhou

ISBN 978-3-642-05011-4 e-ISBN 978-3-642-05012-1
Springer Heidelberg Dordrecht London New York

Library of Congress Control Number: 2009936204

Cover design: Frido Steinen-Broo, EStudio Calamar, Spain

Printed on acid-free paper

Springer is a part of Springer Science+Business Media (www.springer.com)

Preface

Nanostructured materials with designed biofunctions have been bringing rapid and significant changes in materials sciences. Nanostructured Biomaterials provides up-to-date reviews of different routes for the syntheses of new types of such materials and discusses their cutting-edge technological applications. The chemical synthesis and physicochemical preparation of nanosized materials are summarized with particular attention on the self-assembly of specific molecular and nanosized building blocks into functional nanostructures. The reviews mainly focus on potential applications of nanostructured materials in biology and medical sciences. The book is of general interest to a wide community of graduate students and researchers active in chemistry, materials science, engineering, biology, and physics.

Within the last decades, rapid advances in nanotechnology spurred great interest in nanostructured materials. In particular nanostructures with biofunctional properties are most promising, challenging traditional materials in many ways. Meanwhile a large variety of nanostructured artificial biomaterials with tailored morphologies and functionalities have been designed and fabricated.

In this book, we present recent achievements in the synthesis and application of nanostructured biomaterials. We will show our readers the exciting challenges in this unique research area and we hope to convince them of the many new research opportunities.

Silica-based mesoporous nanomaterials show remarkable potential as drug-delivery systems and biosensors. They are reviewed by Yang Yang and Junbai Li in Chapter 1. Natural substances possess sophisticated hierarchal structures. Yuanqing Gu and Jianguo Huang summarize in Chapter 2 how they are utilized as templates and/or scaffolds for the fabrication of nanostructured materials. In Chapter 3, Peiqin Tang and Jingcheng Hao introduce polyoxometalate-based hybrid nano-materials, which are used especially for thin films formed by different deposition techniques. Nanometer-precise coatings of metal oxides on morphologically complex surfaces of natural cellulose substances are addressed in Chapter 4. It is shown how metal oxide, polymer, and protein-immobilized nanomaterials are produced by using "old-fashioned" biocellulose. In the last chapter, Yue Cui, Qiang He and Junbai Li describe functional nanomaterials that are synthesized by employing porous membranes as templates.

The editor thanks the editorial staff, Ms Xiaojia Chen, Mr Jianzhong You and Ms Ge Zhang, for their excellent professional support.

Acknowledgements

The work of Chapter 1 was supported by the National Key Project on Basic Research of China (No. 2009CB930101).

The work of Chapter 2 was supported by the National Key Project on Basic Research of China (No. 2009CB930104).

The work of Chapter 3 was supported by the National Natural Science Foundation of China (Grant No. 20625307) and the National Key Project on Basic Research of China (No. 2009CB930103).

Most of Jianguo Huang's own research works presented here were done in Prof. Toyoki Kunitake's laboratory and under his guidance in RIKEN, Japan. The work of Chapter 4 was supported by the National Key Project on Basic Research of China (No. 2009CB930104).

Junbai Li
Beijing, China
August 2009

Contents

Contributors

Jianguo Huang	Department of Chemistry, Zhejiang University, Hangzhou, Zhejiang, 310027, China
Jingcheng Hao	Key Laboratory for Colloid and Interface Chemistry, Ministry of Education, Shandong University, Jinan, 250100, China
Junbai Li	National Center for Nanoscicence and Technology, Beijing, 100190, China Beijing National Laboratory for Molecular Sciences (BNLMS), International Joint Lab, CAS Key Lab of Colloid and Interface Science, Institute of Chemistry, Chinese Academy of Sciences, Beijing, 100190, China
Peiqin Tang	Key Laboratory for Colloid and Interface Chemistry, Ministry of Education, Shandong University, Jinan, 250100, China
Qiang He	Beijing National Laboratory for Molecular Sciences (BNLMS), International Joint Lab, CAS Key Lab of Colloid and Interface Science, Institute of Chemistry, Chinese Academy of Sciences, Beijing, 100190, China
Yang Yang	National Center for Nanoscicence and Technology, Beijing, 100190, China
Yuanqing Gu	Department of Chemistry, Zhejiang University, Hangzhou, Zhejiang, 310027, China
Yue Cui	Beijing National Laboratory for Molecular Sciences (BNLMS), International Joint Lab, CAS Key Lab of Colloid and Interface Science, Institute of Chemistry, Chinese Academy of Sciences, Beijing, 100190, China

1

Silica-based Nanostructured Porous Biomaterials

Yang Yang[1] and Junbai Li[1,2]

[1]National Center for Nanoscicence and Technology, Beijing, 100190, China.
E-mail: yangyang@nanoctr.cn
[2]Beijing National Laboratory for Molecular Sciences (BNLMS), International Joint Lab, CAS Key Lab of Colloid and Interface Science, Institute of Chemistry, Chinese Academy of Sciences, Beijing, 100190, China. E-mail: jbli@iccas.ac.cn

1.1 Introduction

Recently, the application of nanomaterials in medical and biological fields has become more important. Nanoparticles (NPs) have been used as sensors, fluorescent markers, clinical diagnoses, drug delivery and MRI contrast agents (Lin et al., 2005). Inorganic, porous, ceramic nanoparticles have several advantages in biological applications. They are readily engineered with the desired size, shape, and porosity, and are often inert. The ceramic materials have surfaces with hydroxyl groups, and thus they are always hydrophilic (Paul, Sharma, 2001; Roy et al., 2003; Gemeinhart et al., 2005). Such natural hydrophilicity can decrease oxide particle clearance by the immune system, and thus increases their circulation time in blood (Barbe et al., 2004). Growing interest has recently emerged in utilizing porous ceramic nanomaterials as carriers in biological systems, exploring typical biocompatible ceramic nanoparticles, such as silica, alumina, and titania (Yih, Al-Fandi, 2006).

The International Union of Pure and Applied Chemistry (IUPAC) categorizes porous materials into three classes: microporous (<2 nm), mesoporous (2~50 nm), and macroporous (>50 nm). According to their pore sizes, terms such as porous nanomaterials, nanoporous materials, and nanostructured porous materials have been widely used to cover a variety of porous materials studied under bionanotechnology

(Sing et al., 1985). We will focus on the silica-based nanostructured porous materials with pore sizes ranging from a few nanometers to several tens of nanometers.

1.2 Silica Porous Materials in Drug Release Systems

Controlled drug-delivery systems (DDSs) have been facing a big challenge since the last decades. These silica-based nanostructured materials contain pores to provide spaces in loading drugs. By controlling morphological size and the shape of the material, one can design the required systems for the control of drug delivery.

1.2.1 Conventional Delivery Systems

An important prerequisite for designing an efficient delivery system is the capability to transport the desired guest molecules to the targets and release them in a controlled manner (Jin, Ye, 2007). Some toxic anti-tumor drugs are not expected to release before reaching the targeted cells or tissues. Biodegradable polymer-based drug-delivery systems highly rely on the hydrolysis-induced erosion of the carrier structure (Couvreur et al., 1995). The release of encapsulated compounds usually takes place too quickly as they are dispersed in water. In the process of drug loading, the polymer systems typically require the use of organic solvents, which might lead to the change of the undesirable structure and/or function of the encapsulated molecules. The *in vivo* degradation of synthetic polymers poses toxicity problems (Couvreur et al., 1995). The naturally selected polymers have problems with the monomer purity. Liposomes or micelles suffer from poor chemical stability. Thus the newly designed materials need to overcome the above shortcomings.

1.2.2 Silica Porous Materials for Release Systems

Since MCM-41 was synthesized in the 1990s as a member of the M41S family of molecular sieves (Kresge et al., 1992), the mesoporous silica material had been proposed as a DDS to solve the above mentioned problems. In general, mesoporous materials are derived from molecular assemblies of surfactants as templates during synthesis (Kresge et al., 1992; Huo et al., 1994; Zhao et al., 1998; Sakamoto et al., 2004). After the removal of the surfactants, the silica mesoporous materials are achieved. As drug carriers, they possess the following features:

(1) An ordered pore network and homogeneous size for the purpose of the drug loading;

(2) A high pore volume to host the required amount of drug molecules;

(3) A high surface area with a high potential for drug adsorption;

(4) A silanol-containing functionalized surface allows better control over drug loading and release;

(5) Micro- to mesoporous silicas can selectively host molecules (Vallet-Regí et al., 2007). These unique features make mesoporous materials good candidates for controlled drug-delivery systems, based on the many investigations which have been done in recent years.

1.2.3 Various Mesoporous Silica in Drug Delivery Systems

Various mesoporous silica such as M41S, FSM, TUD, and SBA have been designed into DDSs. MCM-41 is the most frequently used mesoporous silica material based drug carrier. They have the ordered hexagonal molecular sieve with large surface areas (>1000 m^2/g), high pore volumes (>0.7 cm^3/g), and a very uniform pore structure (pore diameter 2~3 nm) (Beck et al., 1992; Kresge et al., 1992). MCM-41 is applied with different pharmaceutical compounds such as ibuprofen (Vallet-Regí et al., 2001; Andersson et al., 2004; Charnay et al., 2004), vancomycin (Lai et al., 2003), model compound fluorescein (Karen, Fisher, 2003), diflunisal and naproxen (Cavallaro et al., 2004), hypocrellin A (Zhang et al., 2004), and aspirin (Zeng et al., 2005). And it is also used by including proteins such as cytochrome C and myoglobin for therapy (Deere et al., 2003). MCM-48, the cubic ordered silica material, has also been utilized for the immobilization of protein (Washmon-Kriel et al., 2000) as well as for the encapsulation of small molecule drugs (Izquierdo-Barba et al., 2005). Kuroda et al. reported that Taxol, an anticancer substance, was adsorbed into FSM-type mesoporous silicas with the pore sizes larger than 1.8 nm, while it was not adsorbed into the channels with the pore sizes less than 1.6 nm, indicating that mesoporous silicas have a molecular sieving property for relatively large molecules. The results obtained indicate the potential application of mesoporous silica as a new synthetic vessel (Hata et al., 1999). Moreover, the siliceous mesoporous material, Technische Universiteit Delft (TUD-1), was also studied as a drug delivery vehicle (Jansen et al., 2001). TUD-1 is one of the new mesoporous materials. TUD-1 is synthesized as siliceous, containing only biocompatible amorphous mesostructured silica. It has a foam-like mesoporous structure, where the mesopores are randomly connected in three dimensions. Heikkila's study proved that the highly accessible mesopore network allowed ibuprofen to be adsorbed into TUD-1 with a very high efficiency and the amount of loaded drug exceeded the reported values for other biocompatible mesoporous silicas such as MCM-41 and MCM-48. The drug dissolution profile of TUD-1 material was found to be much faster and to have more diffusion when compared to the mesoporous MCM-41 material (Heikkila et al., 2007). Another mesostructured silica with 2D hexagonal structures, SBA, was also often used as DDSs. Qu et al. employed MCM-41 and SBA materials with variable pore sizes and morphologies as controlled delivery systems for the water soluble drug captopril. Captopril could be successfully loaded into the channel of mesoporous silica materials. The drug loading and release kinetics was correlated to morphologies and pore sizes of mesoporous silica (Qu et al., 2006). Adsorption experiments carried out with alendronate (Vallet-Regí et al., 2007) (small molecule) and albumin (Manzano et al., 2006) (macromolecule) on SBA-15 indicate that the very high or very low drug molecule/pore size ratios are, in both cases, inadequate for incorporating large amounts of drugs.

1.2.4 Stimuli-responsive Mesoporous Silica for Delivery Systems

It is highly desirable to design delivery systems that can respond to external stimuli and release the guest molecules at specific sites. To achieve this goal, several groups developed a series of stimuli-responsive mesoporous silica delivery systems, including photo-responsive, pH-responsive, thermo-responsive, and enzyme-responsive delivery systems.

1.2.4.1 Photo-responsive System

Fujiwara et al. have developed a photo-responsive release system for direct-drug release applications based on pore-entrance modification with coumarin groups. These groups undergo reversible dimerization upon irradiation with UV light at wavelengths longer than 310 nm, and return to the monomer form by subsequent irradiation at shorter wavelengths. The dimer form of the coumarin, when grafted on the surface of mesoporous silica systems, reduces the effective pore size of the matrix, and subsequently hinders the adsorption of molecules into the pore voids as well as their release from them. Adequate irradiation of the material opens the entrance to the pores and the adsorbed drugs can be released (Mal et al., 2003a; 2003b). Another photo controlled DDS based mesoporous silica is a kind of molecular machine called a “nanoimpeller” developed by Zink group (Angelos et al., 2007a; Lu et al., 2008). It was made by immobilizing an active molecule having photo responsive behaviors such as azobenzene derivatives to the mesostructured silica framework. It is reported that azobenzenes in nanostructured silica will go through *cis*-trans isomerization after continuous illumination at 413 nm (Liu et al., 2003a; Sierocki et al., 2006). The bifunctional strategy was used to attach a small azobenzene to the interiors of the pores templated by the surfactant. This method involved the coupling reaction of the azobenzene with a silane linker followed by co-condensation with the TEOS silica precursor (Liu et al., 2003b). After removing the surfactant, particles contained azobenzenes with one side bonded to the inner pore walls and the other free to undergo reversible isomerization which creates a large amplitude wagging motion capable of functioning as nanoimpellers to release pore contents from the particles (Fig.1.1).

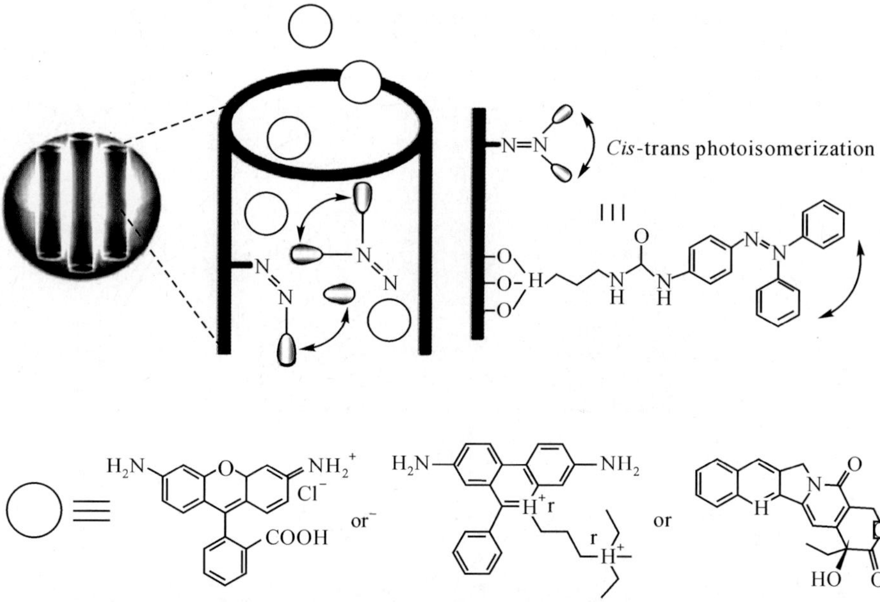

Fig.1.1. Designed pore interiors of the light-activated mesostructured silica (LAMS) nanoparticles functionalized with azobenzene derivatives. Continuous illumination at 413 nm causes a constant *cis*-rans photoisomerization about the N-N bond causing dynamic wagging motion of the azobenzene derivatives and results in the release of the molecules through and out of the mesopores. Copyright (2008), with permission from Wiley

1.2.4.2 pH-responsive System

Ionically controlled nanoscopic molecular gates were also designed by using functionalized mesoporous materials by Marinez-Manez et al. (Casasus et al., 2004). The molecular gate referred to a basic device that modulates the access to a certain site and whose state (opened or closed) can be controlled by certain external stimuli. In their study, the functionalized mesoporous silica materials were synthesized so that the polyamines groups were at the external surface while the mercaptothiol groups remained inside the mesopores (Fig.1.2). They introduced a pH-controlled gate mechanism which comes from hydrogen-bonding interactions between amines (open state) and coulombic repulsion between ammonium groups (closed state). They also verified pH-controlled and anion-controlled mechanisms by using a colorimetric reaction consisting of the selective bleaching of a blue squaraine dye by reaction with the mercaptopropyl groups. Xiao and co-workers also designed pH-responsive carriers in which polycations are grafted to anionic, carboxylic acid modified SBA-15 by ionic interactions (Yang et al., 2005). Drug molecules such as vancomycin can be stored and released from the pore voids of SBA-15 by changing pH at will. In this system, the polycations act as closed gates to store the drug within the mesopores (Fig.1.3). When the ionized carboxylic acid groups are protonated in response to a change in pH, the polycations are detached from the surface and the drug is released from the mesopores.

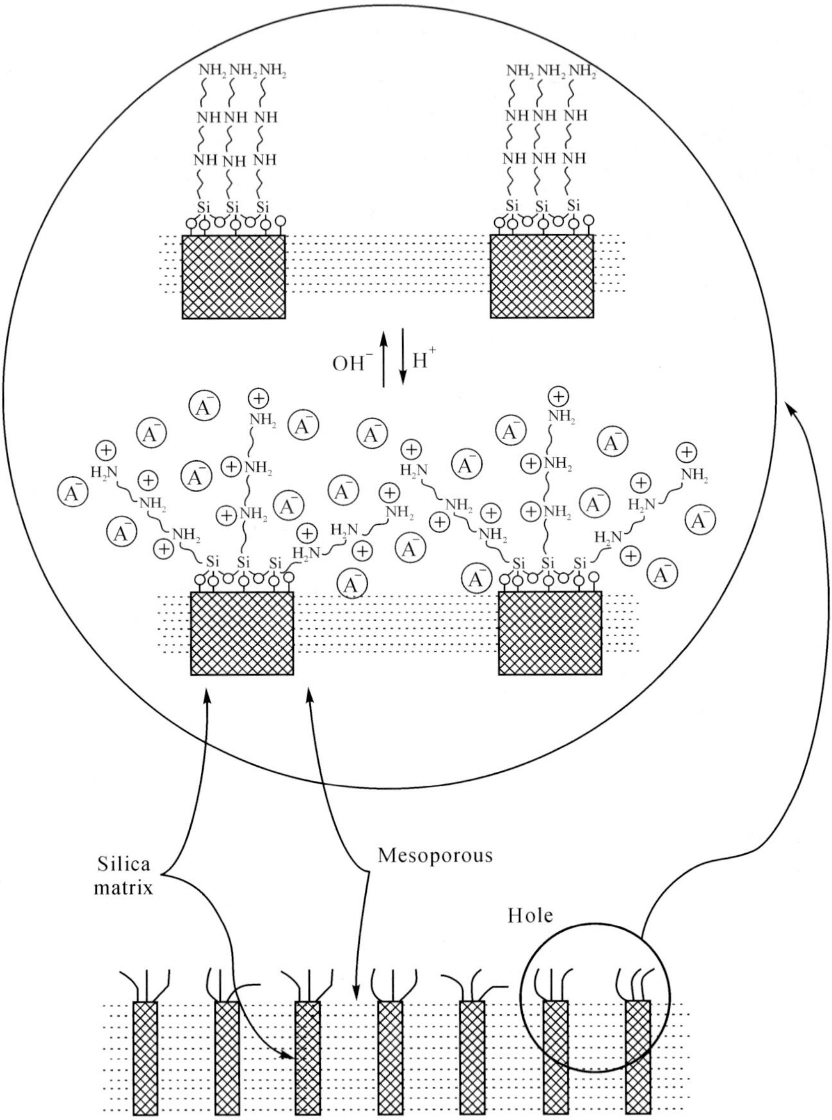

Fig.1.2. Representation of solid Si with a scheme of the ionically controlled nanoscopic "Molecular Gate" mechanism. Copyright (2004), with permission from the American Chemical Society

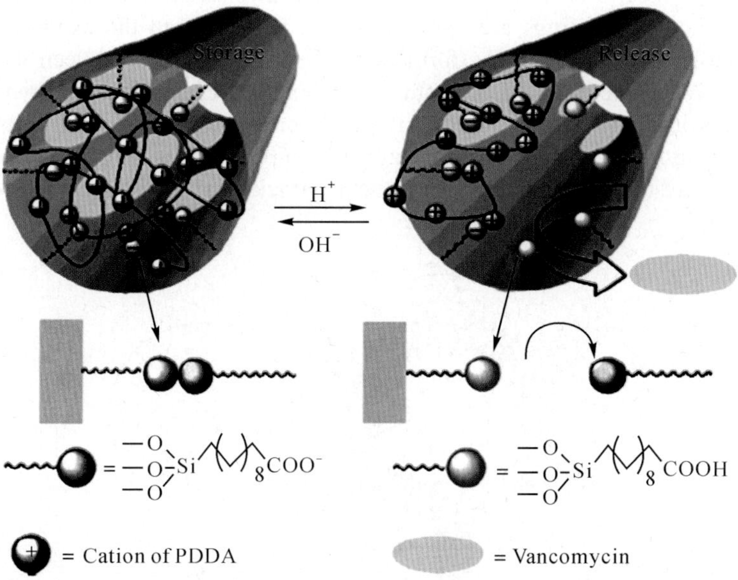

Fig.1.3. Schematic representation of pH-responsive storage-release drug delivery system. This pH-controlled system is based on the interaction between negative carboxylic acid modified SBA-15 silica rods with polycations. Copyright (2005), with permission from the American Chemical Society

1.2.4.3 Thermo-responsive System

Thermosensitive polymer, such as poly (*N*-isopropylacrylamide) (PNIPA), has often been used as a drug delivery material because of its lower critical solution temperature (LCST) at about 32 °C which is close to physiological temperature (Pelton et al., 1989; Pelton, 2000; Li et al., 2007). PNIPA undergoes a thermo-induced conformational change from the swelled, hydrophilic state to the shrunken, hydrophobic state above LCST in water. López and co-workers prepared PNIPA-grafted MCM particles for controlling molecular transportation (Fu et al., 2003; 2007). In their work, the porous network of silica was modified by PNIPA by atom transfer radical polymerization (ATRP). At lower temperature, *e.g.* room temperature, the PNIPA is hydrated and extended, and inhibits transport of solutes; at higher temperature, *e.g.* 50 °C, it is hydrophobic and is collapsed within the pore network, thus allowing solute diffusion. Uptake and release of fluorescent dyes from the particles were verified by several characterization methods. Zhu et al. also fabricated a site-selective controlled delivery system for controlled ibuprofen (IBU) release through the *in situ* assembly of thermo-responsive ordered SBA-15 and magnetic particles (Zhu et al., 2007). The approach is based on the formation of ordered mesoporous silica with magnetic particles formed from $Fe(CO)_5$ via the surfactant-template sol–gel method and control of transport through polymerization of *N*-isopropylacrylamide inside the pores. The system combines the advantages of mesoporous silica, thermosensitive PNIPA multilayers and magnetic particles. At low temperature, the drugs are confined to the pores due to the expansion of the PNIPA molecular chain and the formation of hydrogen bonds between the PNIPA and IBU. When increasing temperature, the polymer chains become hydrophobic and swell within the pore network, driving the drug molecules to be released from the pores (Fig.1.4). The materials they prepared can be used as temperature controlled drug release systems by inducing the magnetic particles and thermosensitive polymer.

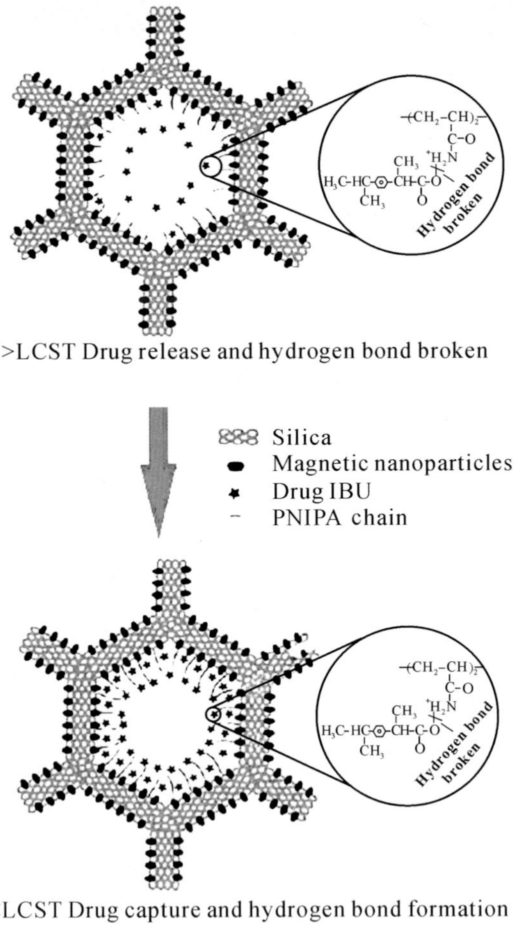

Fig.1.4. Schematic representation of the stimuli-responsive delivery system based on SBA-15 with magnetic particles formed as the target label (LCST: lower critical solution temperature). Copyright (2007), with permission from Wiley

1.3 Mesoporous Silica Nanoparticles

After the discovery of highly ordered mesoporous silica materials by scientists at the Mobil Corporation in 1992, significant research efforts have been under way to achieve control over the characteristics of mesoporous silica with special emphasis on pore size and morphology. Through this vast research, new families of mesoporous silica materials, such as SBA (Zhao et al., 1998), MSU (Bagshaw et al., 1995), and FSM (Inagaki et al., 1993), were developed with characteristic porosities and particle shapes.

1.3.1 MSNs for Biological Applications

Most of those materials consisted of particles with sizes in the micrometer scale and thus would be poorly dispersible. Recently, various mesoporous silica nanoparticles (MSNs) with well-defined and controllable particle morphology were developed in Lin's and other research groups in the pursuit of biocompatible materials to be used in controlled release and drug delivery systems (Trewyn et al., 2007; Yang et al., 2008; Zhang et al., 2008). These mesoporous silica materials, all nano-sized, are more valuable in biological application and drug delivery.

1.3.2 Non-functionalized MSNs in Drug Release Systems

In 2001, Maria's group loaded ibuprofen into MCM-41 materials with different pore sizes and studied the drug release in a simulated body fluid (Vallet-Regí et al., 2001). The results showed that the MCM-41 type mesoporous structure, which with channel-like pores packed in a hexagonal fashion, was able to load large quantities of drug molecules and release them over a relatively long period of time. However, no significant difference was observed between the release profiles of materials with different pore sizes. To investigate the impact of the pore and particle morphology of mesoporous silica materials to the controlled release properties, a series of room temperature ionic liquids (RTILs) containing MSN materials with various particle morphologies, such as spheres, ellipsoids, rods, and tubes, etc., was prepared by Lin's group (Trewyn et al., 2004). It also can be seen that the pore morphologies were tuned from MCM-41-type hexagonal mesopores to rotational helical channels, and to wormhole-like porous structures by changing the ionic liquid template (Fig.1.5). The controlled release of the pore-encapsulated ionic liquids was also investigated by measuring the antibacterial effect on Escherichia coli. All the results indicated that the spherical, hexagonal MSN exhibited a superior antibacterial activity to that of the tubular, wormhole MSN, which was attributed to the fact that the rate of RTIL release via diffusion from the parallel hexagonal channels would be faster than that of the disordered wormhole pores. This work categorically demonstrates the essential role of the morphology of mesoporous silica nanomaterials on controlled release behavior.

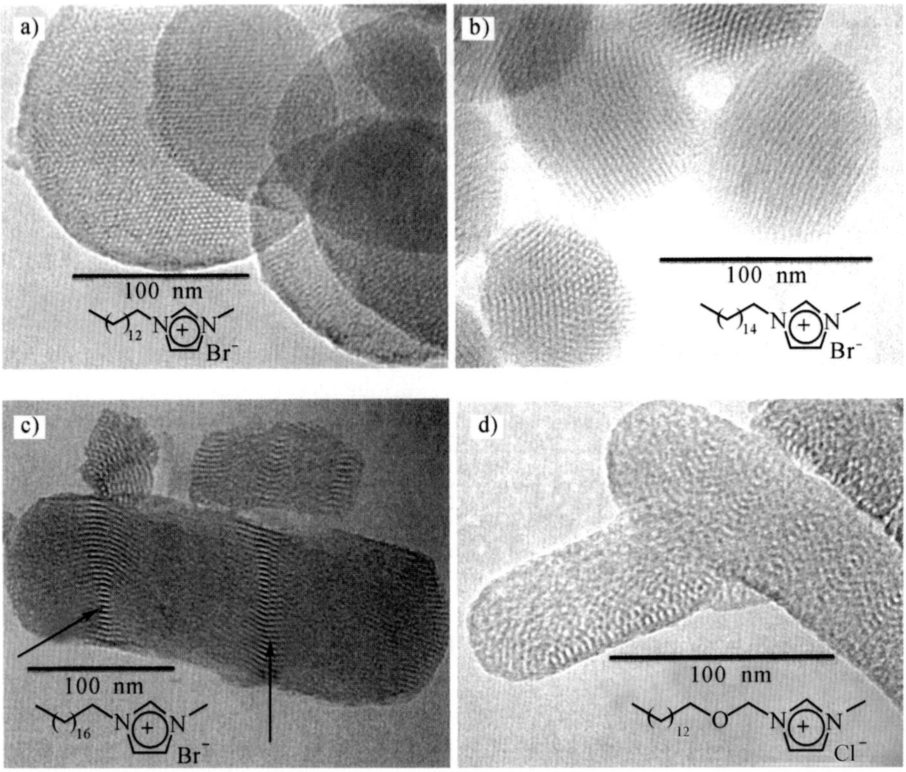

Fig.1.5. TEM of MSN with different morphologies prepared by Lin's group. a) Spheres, b) ellipsoids, c) rods, and d) tubes. Copyright (2004), with permission from the American Chemical Society

Qiao et al. also prepared different helical morphological MSN particles by the co-condensation of TEOS and hydrophobic organoalkoxysilane, such as 3-mercaptopropyltrimethoxysilane (MPTS), using achiral surfactants as templates (Zhang et al., 2008). They synthesized surfactant-extracted samples by using either $C_{16}TAB$ or $C_{18}TAB$ as a template. The morphology and pitch of helical mesostructured silica can be controlled by simply varying the amount of added organoalkoxysilane MPTS. When no MPTS is added, the product is spherical with a diameter of 60~80 nm; when a small amount of MPTS (0.02~0.10 g) is added, the materials change into helical mesostructured rods (Fig.1.6). They also investigated the drug release rate in mesostructured silica samples with different morphology and helicity but similar pore sizes. The results indicated that the drug release rate can be controlled by the morphology and helicity of the helical mesostructured silica.

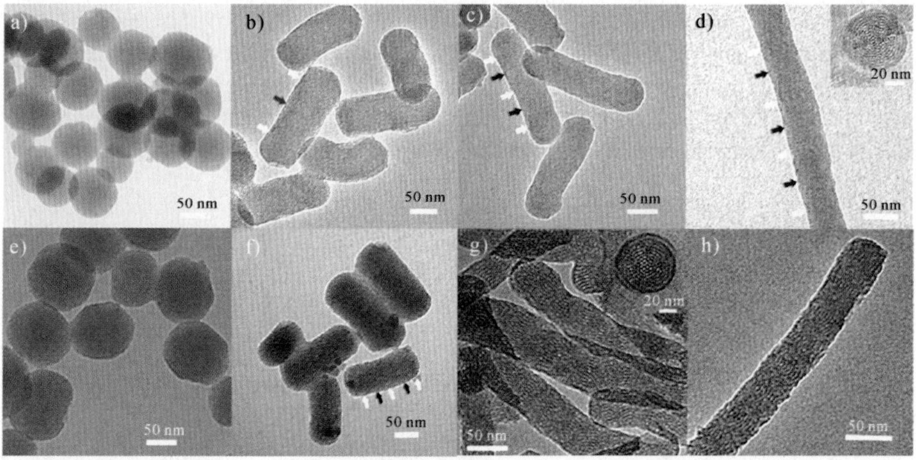

Fig.1.6. TEM images of the samples synthesized with different amounts of MPTS using C_{16}TAB: a) C_{16}-0.0, b) MPTS-C_{16}-0.05, c) MPTS-C_{16}-0.08, d) MPTS-C_{16}-0.10 (inset: cross-section images); TEM images of the samples synthesized with different amounts of MPTS using C_{18}TAB: e) C_{18}-0.0, f) MPTS-C_{18}-0.02, g) MPTS-C_{18}-0.05 (inset: cross-section images), h) MPTS-C_{18}-0.10. Copyright (2008), with permission from Wiley

1.3.3 Inorganic Nanocrystals Capped MSNs

The unique hexagonally ordered pore structures of the MSNs offer an important advantage for drug delivery over those of porous hollow silica nanoparticles. It is because the pores of MSN are independent parallel channels without any interconnection and each pore only has two openings. A zero premature release system can be generated in a situation of imperfect capping, as long as both of the openings of a single mesopore are capped. Lin and co-workers developed a series of inorganic nanocrystals capped MSNs delivery systems (Trewyn et al., 2007). For example, they used CdS nanoparticles as the pore caps by attaching them to the entrance of the pores of the mesoporous MCM-41 nanospheres, as depicted in Fig.1.7 (Lai et al., 2003). The mesopores loaded with guest molecules were capped by CdS nanoparticles via a chemically cleavable disulfide linkage to the MSN surface. Being physically blocked, guest molecules were unable to leach out from the MSN host, thus preventing any premature release. The guest molecules could be released by exposing the capped MSNs to a chemical reagent that can remove the nanoparticle caps. In a similar way, they also introduced site-directing capability to the MSN-based delivery system. Mesoporous silica nanorods capped with super-paramagnetic iron oxide nanoparticles were fabricated (Giri et al., 2005). By using cell-produced antioxidants as triggers in the presence of an external magnetic field, guest molecules that are smaller than 3 nm, such as fluorescein, can be encapsulated and released from the magnet-MSN delivery system. Furthermore, the biocompatibility of this system was also investigated. HeLa cells were incubated in the presence of magnet-MSN to allow for the internalization of this material. The internalized magnet-MSN cells were isolated, and the site direction was demonstrated by using a magnet to move these internalized magnet-MSN HeLa cells across a cuvette. The results of confocal fluorescence microscopy indicated that the magnet-MSNs were indeed internalized and had released fluorescein from the pores (Fig.1.8).

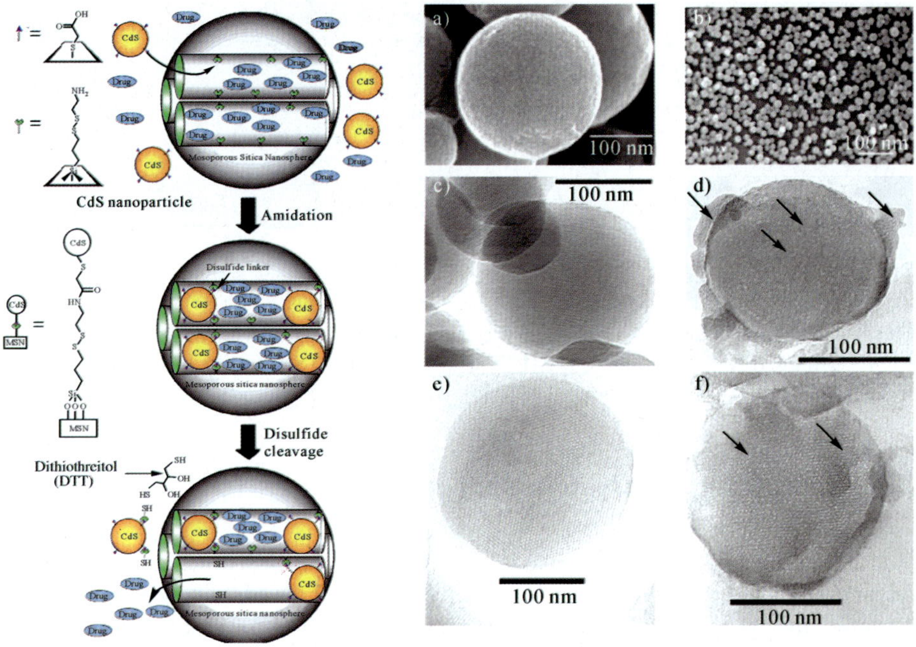

Fig.1.7. Schematic representation of the drug/neurotransmitter delivery system based on the CdS nanoparticle-capped MCM-41 type mesoporous silica nanospheres. The controlled-release mechanism of the system is based on chemical reduction of the disulfide linkage between the CdS caps and the MSN hosts. SEM (a, b) and TEM micrographs of the linker-MSN (c, e). The TEM micrographs (d, f) of the CdS-capped MSN exhibit aggregations of CdS nanoparticles on the exterior surface of MSN material represented by dots in the areas indicated by black arrows in d). Copyright (2003), with permission from the American Chemical Society

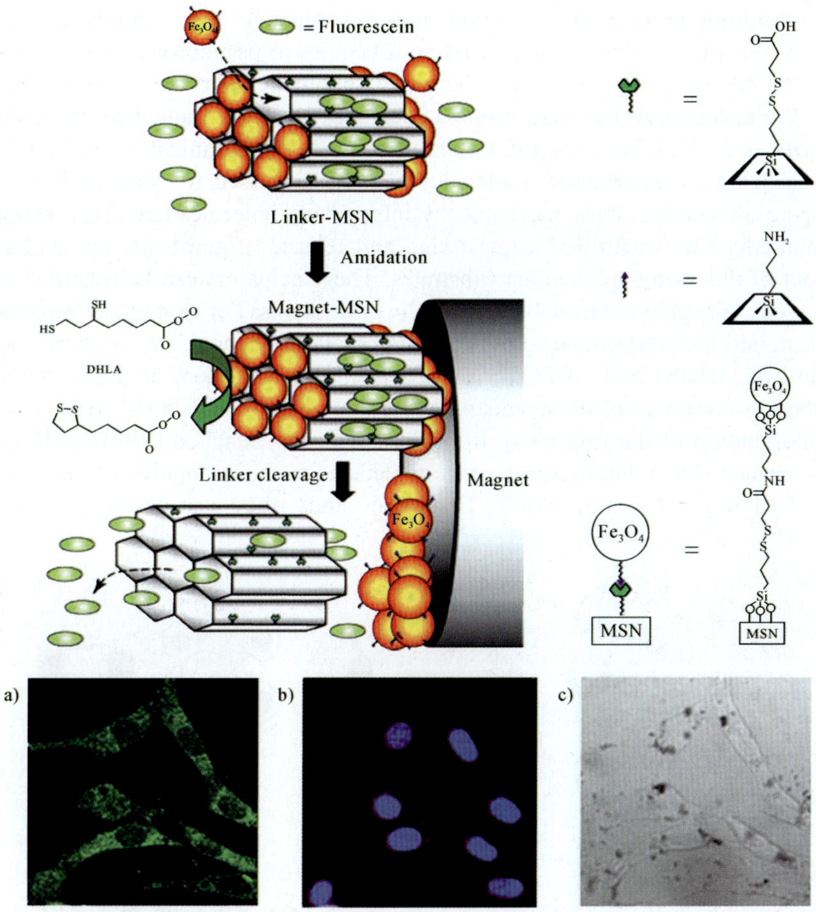

Fig.1.8. Schematic representation (above) of the superparamagnetic iron oxide nanoparticle-capped mesoporous silica nanorod material (magnet-MSN). Panels a), b), and c) are fluorescence confocal micrographs of HeLa cells after 10 h incubation with Fe_3O_4-capped fluorescein-loaded MSNs: a) cells excited at 494 nm; b) cells excited in the UV region; c) a pseudo-brightfield image, where dark aggregations of magnet-MSNs can be clearly observed. Copyright (2005), with permission from Wiley

1.3.4 The "Nanocalves" on the Surface of MSNs

Besides the nanoparticle-capped MSN materials mentioned above, Zink and co-workers developed molecular machines supported on the surface of MSNs for controlling release applications (Hernandez et al., 2004; Nguyen et al., 2005; Angelos et al., 2007; Saha et al., 2007; Patel et al., 2008). The machines are designed to block the nanopores in one configuration to trap guest molecules, but unblock the nanopores in another configuration. Therefore, the guest molecules can be released on command.

The switching motion of supramolecular machines is also called nanovalves. Nanovalves are machines which based on rotaxanes or pseudorotaxanes, and can be operated mechanically by redox chemistry, photochemistry or electrochemistry. After the pseudorotaxanes and bistable rotaxanes have been attached covalently to the orifices of the silica nanopores, stimuli-controlled mechanical movement within the mechanically interlocked molecules can be harnessed to close and open the nanopores. Therefore, these mechanically interlocked molecules have been employed as nanovalves for controlled sequestering and release of guest dye molecules into and out of the mesoporous silica substrates. These actuators can be regarded as the prototypes of highly controllable drug-delivery systems. For example, Hernandez et al. designed the redox-controllable supramolecular nanovalves system. In this system, the DNPD and $CBPQT^{4+}$ pseudorotaxane were taken as gatekeepers and attached to mesoporous silica substrates. Reduction of the $CBPQT^{4+}$ ring results in the dissociation of the ring away from the covalently attached DNPD stalk on the silica surface. Then, luminescent probe molecules in the nanopores can be released (Fig.1.9) (Hernandez et al., 2004). In another study, they also designed a snap-top

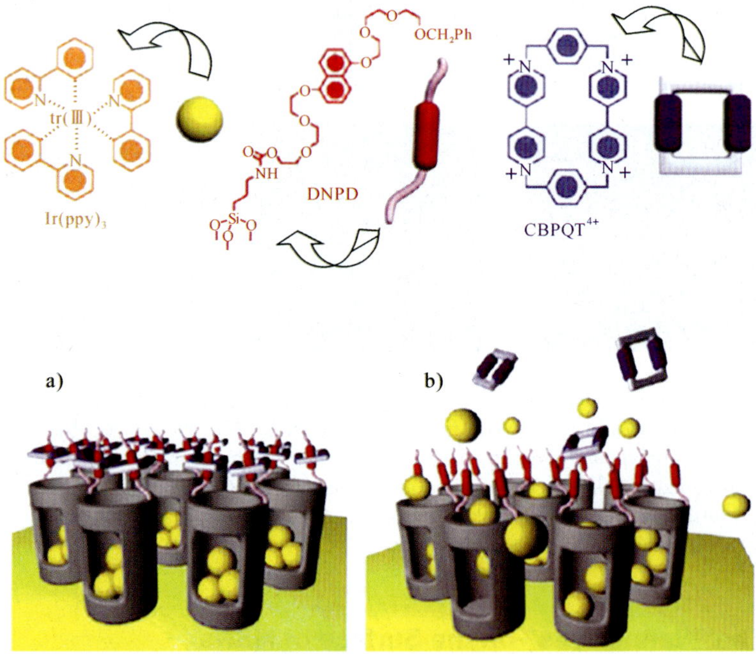

Fig.1.9. Graphical representations of operation of nanovalves. a) The orifices of the nanopores (diameter 2 nm) are covered with pseudorotaxanes (formed between DNPD and $CBPQT^{4+}$) which trap the luminescent $Ir(ppy)_3$ molecules inside the nanopores; b) Upon their reduction, the $CBPQT^{2+}$ bisradical dications are released and allow the $Ir(ppy)_3$ to escape. Copyright (2004), with permission from the American Chemical Society

system in which rotaxanes were assembled on the surface of MSNs to encapsulate or release guest molecules on command. In this snap-top system, a triethylene glycol thread encircled by a cyclodextrin macrocycle is played to a cleavable stopper. The bulky macrocycle traps guest molecules in the pores and the contents can be released when the stopper was cleaved (Fig.1.10) (Patel et al., 2008).

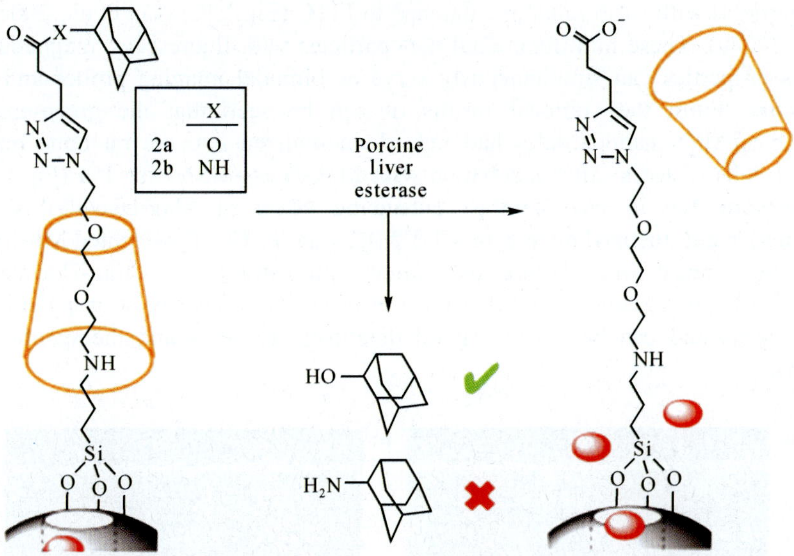

Fig.1.10. Snap-tops attached to the surface of mesoporous silica nanoparticles are able to store guest molecules within the pores while intact. Guest molecules are released upon selective cleavage of esterlinked adamantyl stoppers by porcine liver esterase. Copyright (2008), with permission from the American Chemical Society

1.3.5 MSNs as Biomarkers

At present, fluorescent nanomaterials have been used as labels in biological and medical applications for imaging (Larson et al., 2003; Kim et al., 2004) and diagnostic purposes (Michalet et al., 2005; Chattopadhyay et al., 2006). However, the advent of silica-based nanomaterials has revolutionized this field in both diagnosis (Alivisatos, 2001) and therapy (Akerman et al., 2002; Holm et al., 2002) owing to their robustness, safety, and multimodality (Ferrari, 2005; Kumar et al., 2008). Furthermore, MSNs have other unique properties, such as rigid structure, large pore volume, uniform pore size, great surface-modification capability, and good biocompatibility which make them more suitable as biomarkers. Some researchers have reported MSNs can act as biomarkers, such as in cell imaging agents and magnetic resonance imaging (MRI) agents in biosystems. For example, Mou and co-workers synthesized FITC functionalized MSN materials and demonstrated the cell labeling capability (Lin et al., 2005). These FITC functionalized nanoparticles with the size of 110 nm were shown to be internalized into fibroblast cells, and to accumulate in cytoplasm. The mesoporous silica nanoparticles appear to have no apparent cytotoxic effects on the fibroblast cells. The cellular internalization seems generic for various cells (Fig.1.11). They also prepared composite nanomaterials by fusing of amorphous silica shells of Fe_3O_4@SiO_2 nanoparticles with MSNs that are attached to FITC (Fig.1.12) (Lin et al., 2006; Wu et al., 2008a). These multifunctional nanoparticles with fluorescent, magnetic, and porous properties can simultaneously serve as bimodal imaging probes and drug reservoirs. From the confocal results, it can be seen that the green-emitting Mag-Dye@MSN nanoparticles had indeed internalized into the rat bone marrow stromal cells (rMSCs) after incubation with Mag-Dye@MSNs for 1 h (Fig.1.13a). Furthermore, the *in vivo* contrast enhancing effect of Mag-Dye@MSNs was evaluated in anesthetized mice with a 7-T MRI system. The T_2-weighted MR images show the mouse liver before and after administration of Mag-Dye@MSNs (Fig.1.13b). The results proved that most Mag-Dye@MSNs can be trapped by the RES organs and can be used as good diagnostic reagents and therapeutic drug carriers.

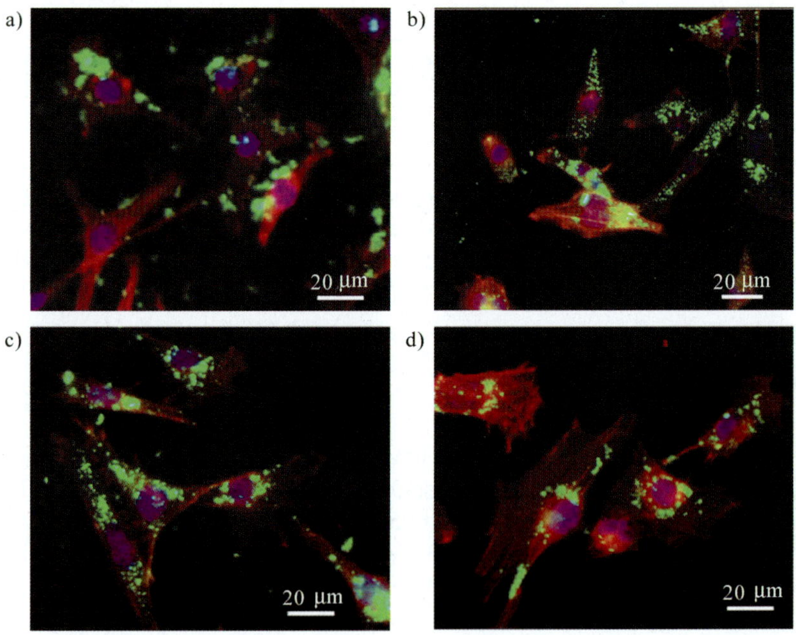

Fig.1.11. Time-course confocal images of 3T3-L1 cells labeled with FITC-MSNs (green emitting). Cell skeleton was stained with rhodamine phalloidin (red), and cell nucleus with DAPI (purple). Cells were incubated with FITC-MSNs for 1 h, washed, and further incubated in particle free medium for a) 0 h, b) 1 d, c) 3 d, and d) 5 d. Copyright (2005), with permission from the American Chemical Society

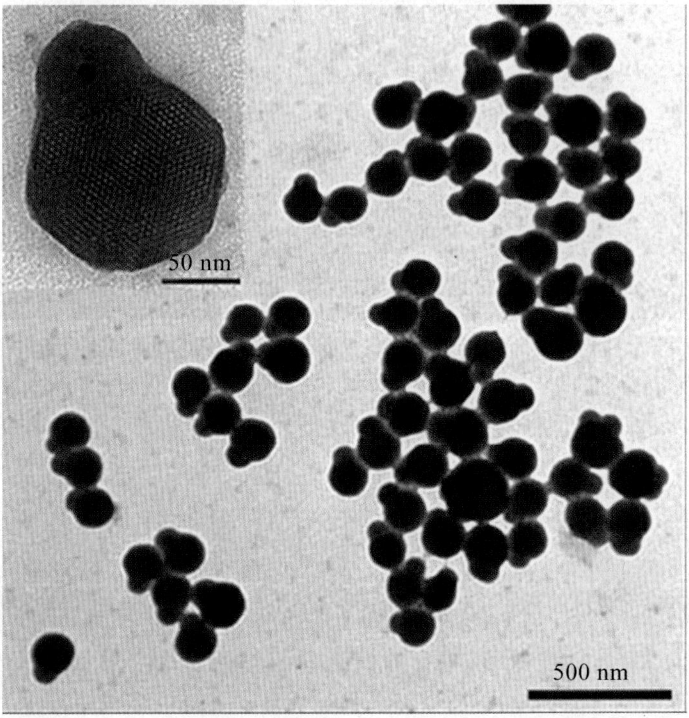

Fig.1.12. TEM image of Mag-Dye@MSNs. The high-magnification TEM image (inset) shows the well ordered mesoporous and magnetic parts. Copyright (2008), with permission from Wiley

Hyeon et al. also synthesized discrete, monodisperse, precisely sized core-shell mesoporous silica NPs smaller than 100 nm by using single Fe_3O_4 nanocrystals as cores (designated as Fe_3O_4@$mSiO_2$) (Kim et al., 2008). They also demonstrated the multifunctional bio-applications of the core-shell NPs for simultaneous magnetic resonance (MR) and fluorescence imaging, and for drug delivery. From the TEM images, they obtained Fe_3O_4@$mSiO_2$ particles which are very uniform and coated wormhole-like mesoporous silica shells. Hollow mesoporous silica nanoparticles were also fabricated by the removal of the Fe_3O_4 nanocrystals (Fig.1.14). For biomedical applications, they modified the surface of the NPs with PEG (Fe_3O_4@$mSiO_2$-PEG) to render them biocompatible by preventing the nonspecific adsorption of proteins to the NPs. When the MCF-7 cells or the nude mice bearing tumors were injected with Fe_3O_4@$mSiO_2$-PEG nanoparticles, both the fluorescence images and the T_2-weighted MR images confirmed accumulation of the materials into the tumor site. The research indicated that the materials are more appropriate *in vivo* applications because of their small sizes and high monodispersible characteristics (Fig.1.15).

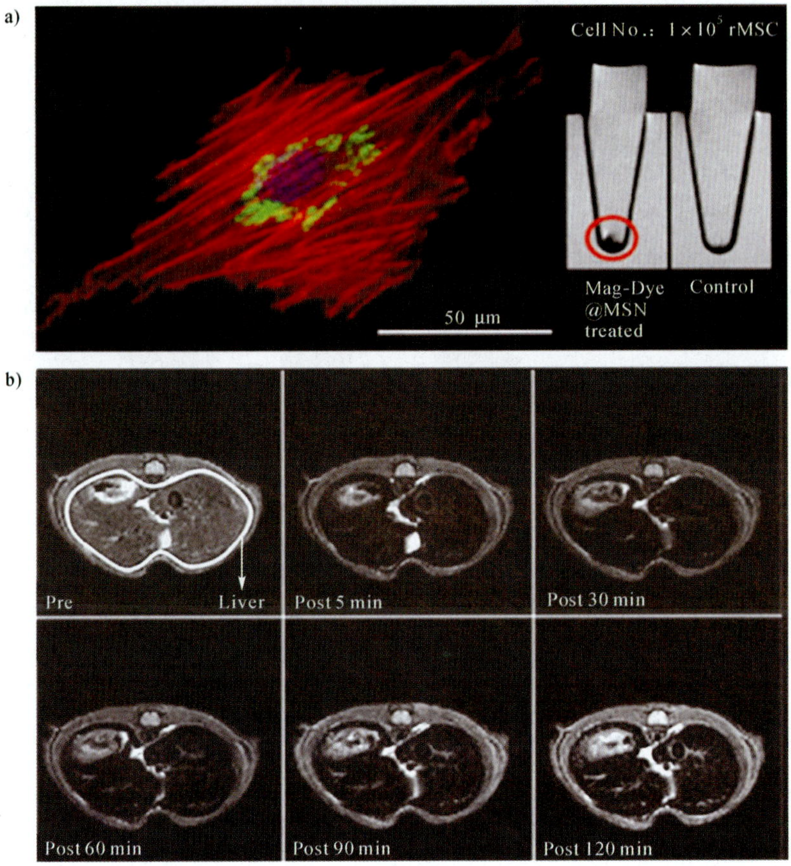

Fig.1.13. a) Merged confocal image of rMSCs that had been allowed to grow in regular growth medium, overnight, after 1 h of incubation with Mag-Dye@MSNs. The cytoskeleton was stained with rhodamine phalloidin (red) and the cell nucleus with DAPI (blue). b) *in vivo* T_2-weighted MR images before (pre) and after (post) administration of 5 mg Fe per kg body weight of Mag-Dye@MSNs (post 5, 30, 60, 90, and 120 min). Liver images showed time-dependent darkening in MR images. Copyright (2008), with permission from Wiley

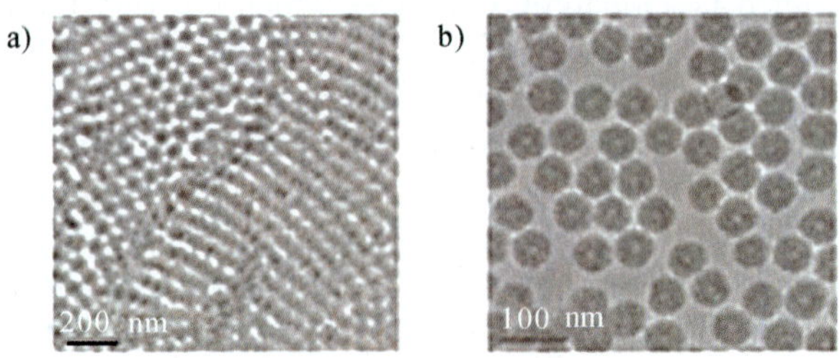

Fig.1.14. TEM images of core-shell and hollow mesoporous silica NPs. a) 53 nm Fe_3O_4@$mSiO_2$ with 15 nm core; b) hollow mesoporous silica NPs from a). Copyright (2008), with permission from Wiley

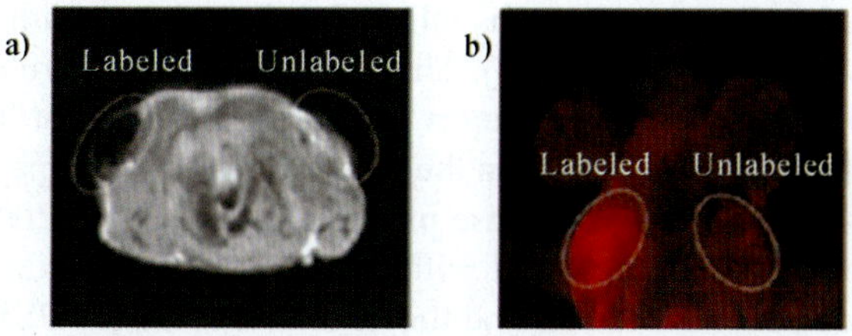

Fig.1.15. *in vivo* multimodal imaging using Fe_3O_4@$mSiO_2$. a) *in vivo* T_2-weighted MR, and b) fluorescence images of subcutaneously injected MCF-7 cells labeled with Fe_3O_4@$mSiO_2$(R) (10 μg Fe mL^{-1}) and control MCF-7 cells without labeling into each dorsal shoulder of a nude mouse. Copyright (2008), with permission from Wiley

1.4 Polymer Coated MSNs

Organic/inorganic hybrid nanoparticles have attracted ever increasing attention in the past decade due to their fascinating optical, electronic, magnetic, and catalytic properties (Caruso, 2001; Wu et al., 2008b; Zou et al., 2008). The hybrid nanoparticles we discussed here mainly refer to MSN cores coated with polymer shell materials. Such materials not only provide porous cores which can act as suitable reservoirs, but also possess functional polymer shells which can improve the surface character of MSN and control the guest molecules release. Therefore, such MSN-polymer core-shell materials are more valuable in biological systems for DDSs or as biosensors. The organic polymer shell mainly determines the surface chemical properties of nanoparticles and their responsiveness to external stimuli, whereas the physical properties of nanoparticles are governed by both the size and shape of MSN cores and the surrounding polymer layer. Polymer coated MSN nanoparticles can be prepared by either physical adsorption or covalent grafting techniques including "grafting-to" or "grafting-from" methods.

1.4.1 Polymer Coated MSNs through Physical Adsorption

Layer-by-layer (LbL) is a famous method to construct multi-membranes on all kinds of substrate, based on electrostatic or other molecular forces (Decher, 1997; Dai et al., 2001). Since the technique was first developed by Decher and others, it has been extensively applied in preparing various materials (Decher et al., 1992; Caruso et al., 2001; Koktysh et al., 2002; Ge et al., 2003; Zheng et al., 2004). It is also a typical approach to modify colloidal particles at a micronanoscale. It is reported that the surface of mesoporous silica is negatively charged above the isoelectric point (pH 2~3) (Kosmulski, 2004) which favors a first-layer coating of the positively charged polyelectrolyte, followed by the negatively charged polyelectrolyte. Thus, polyelectrolyte multilayer coatings on mesoporous silica spheres can be prepared. Shi group synthesized hollow mesoporous silica (HMS) spheres with a 3D mesopore-network shell (Li et al., 2003a; 2003b). They also investigated drug storage capacity and release properties of these spheres, which indicated that they could store significantly more drug molecules than the conventional MCM-type mesoporous silica materials, and also have sustained-release properties (Zhu et al., 2005). Furthermore, the HMS spheres which were coated with polyelectrolyte multilayers consisted of positively charged polycation PAH and the negatively charged PSS by LbL method (Fig.1.16). The materials can be used as containers for drug molecules and the polyelectrolyte multilayer coatings as a pH-responsive switch. This system they designed can not only increase the drug-storage capacity, but also achieve stimuli-responsive controlled release of drug molecules and further enhance the mechanical strength of the polyelectrolyte multilayers to prevent them from breaking.

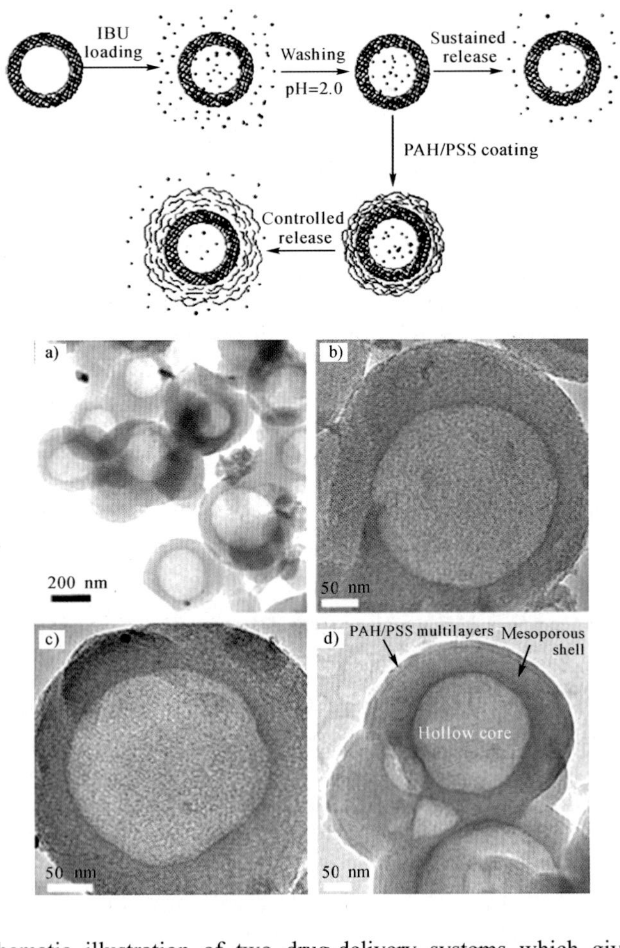

Fig.1.16. Schematic illustration of two drug-delivery systems which give the different controlled-release patterns. TEM micrographs of HMS (a, b), the IBU-HMS system c), and the IBU-HMS@PEM system d). Copyright (2005), with permission from Wiley

The other main property of polymer coated MSN particles is that polymer can introduce some functional biomolecules onto the surface of MSN and result in further bioapplications. For example, Yu et al. also prepared polymer coated HMS particles by combining the template method and LbL technique. The organic dye fluorescein diacetate (FDA) molecules were loaded onto HMS particles followed by polyelectrolyte encapsulation by the LbL method. The polyelectrolyte multilayers provide a suitable interface for further attachment of antibodies (Figs.1.17 and 1.18) (Cai et al., 2008). The labeling systems were stimuli responsive to the addition of concentrated NaOH with the loaded dye molecules being released and detected in a well-controlled manner. When applied in sandwich immunoassays, results indicated that the biolabels were immuno-active and generated a higher optimal signal than the conventional dye labeled antibody system.

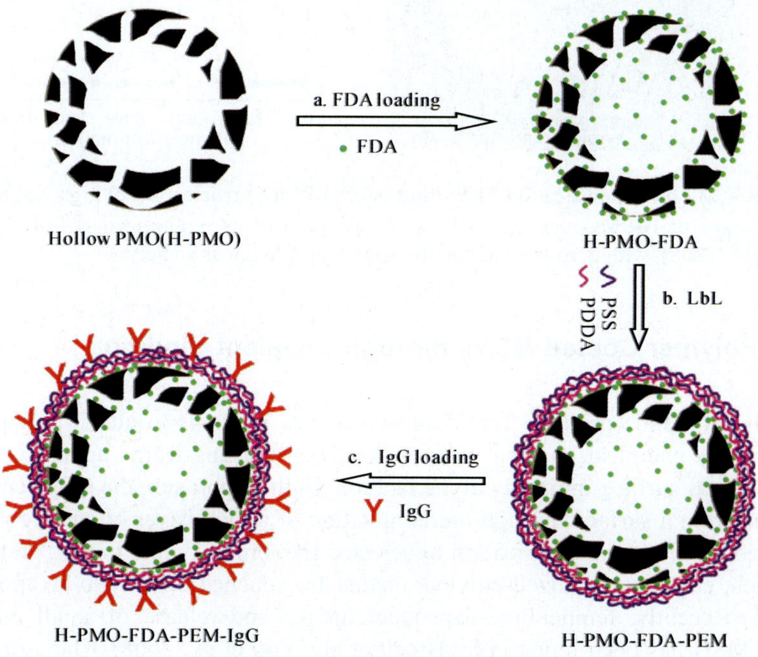

Fig.1.17. Schematic illustration of the preparation of the antibody coated H-PMO-FDA-PEM biolabels: FDA loading into H-PMO particles; LbL assembly of oppositely charged polyelectrolytes onto H-PMO-FDA particles; further coating of IgG resulting in the formation of H-PMO biolabels. Copyright (2008), with permission from the American Chemical Society

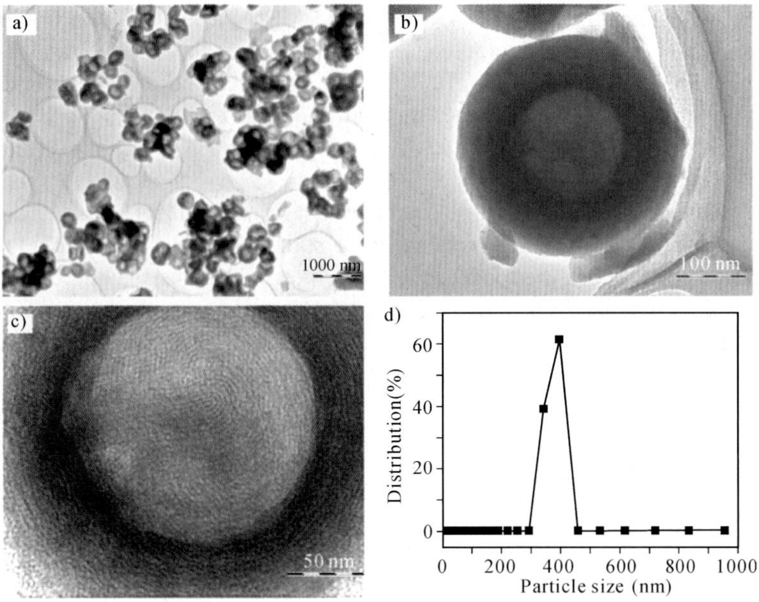

Fig.1.18. a) Low magnification TEM images of H-PMO particles; b), c) high magnification TEM images of H-PMO particles; d) size distribution of as-synthesized H-PMO particles. Copyright (2008), with permission from the American Chemical Society

1.4.2 Polymer Coated MSNs through Covalent Binding

"Grafting-to" and "grafting-from" are two typical methods to attach polymer onto substrate by chemical covalent interaction. The "grafting-from" approach is often referred to as surface-initiated polymerization. In this approach, the polymer chain is initiated from a surface through immobilization of a monolayer of surface initiators followed by *in situ* polymerization of selected monomers, while in the "grafting-to" approach, end-functionalized polymer chains are attached directly to an appropriate surface. Recently, temperature-dependent uptake and release of small molecules within MSN has been reported by Brock et al. (You et al., 2008). They grafted the thermosensitive polymer, PNIPAM, to the surface of the preformed, thiol-functionalized MSN nanoparticles by the "grafting-to" method. The resulting nanoparticle-polymer composites show uptake and release of fluorescein at room temperature and a low level of leakage at 38 °C. At the same time, our group used MSN as a template to prepare MSN@PNIPAM core-shell structures by surface-initiated atom-transfer radical polymerization (ATRP) and the "grafting-from" method (Yang et al., 2008). The as-made core-shell material has also both a mesoporous silica core and a thermosensitive PNIPAM shell (Fig.1.19). The ATRP technique can incorporate more condensed polymers onto the substrate surfaces compared to the "grafting-to" method. The results also indicated that thermosensitive

PNIPAM layer on the particle surface was uniform with controllable thickness, while the inner channels remained. The measured LCST of MSN@PNIPAM material was about 32 °C, consistent with pure PNIPAM in water. In our research, a model drug molecule, FITC, can be entrapped by the stretched polymer chains at low temperature and be locked inside MSN at 40 °C. What's more, the composite materials can be easily internalized into MCF-7 cells which make it a promising material for application in intercellular imaging (Fig.1.20) .

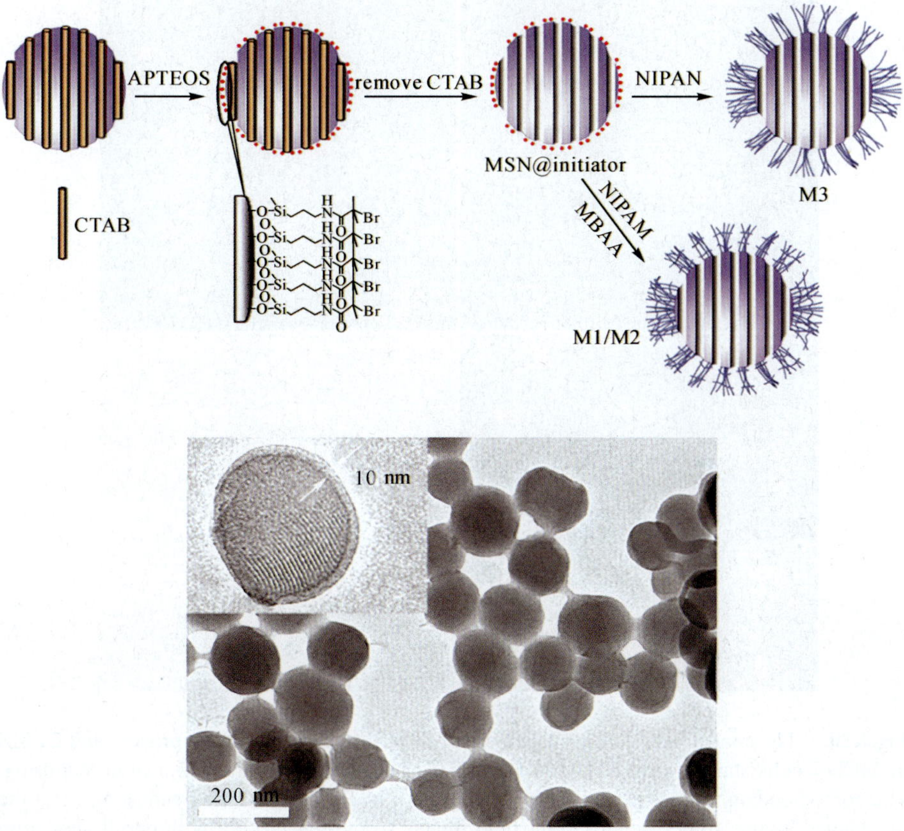

Fig.1.19. A schematic illustration of the surface functionalization for MSN and the fabrication of MSN@polymer and TEM of MSN@PNIPAM through "graft-from" method. Copyright (2008), with permission from the Royal Society of Chemistry

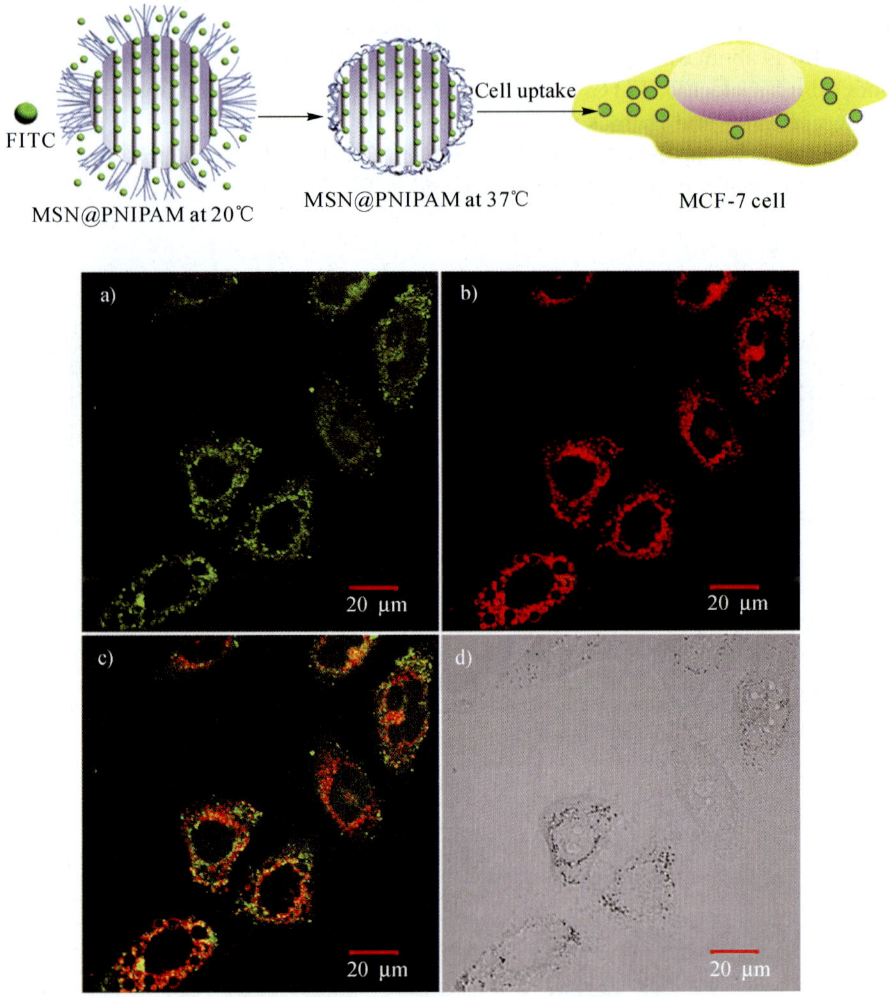

Fig.1.20. The controlled release mechanism of FITC-M3 and cell uptake process and CLSM of MCF-7 cells stained with FM 4-64 and FITC-M3 suspensions after 12 h of co-culturing. The corresponding images: a) the FITC-M3 (green); b) FM 4-64-labeled endosomes ; c) the overlapped image; d)the pseudo-bright field image. Copyright (2008), with permission from the Royal Society of Chemistry

1.5 Summary

In this chapter, we reviewed the recent research progress on the design of functional mesoporous silica materials from several research groups. The synthesis of different functional mesoporous nanomaterials is still in progress. One thing is clear, that every newly constructed material has to be satisfied with the specific demands of biological systems in their own way. To prepare the multifunctional nanostructured drug carriers, it is still necessary to develop new methods which are simple, inexpensive, and environment-friendly. If it becomes a reality, the mesoporous silica drug carriers will play an important role in biotechnology.

References

Akerman ME, Chan WC, Laakkonen P, Bhatia SN, Ruoslahti E (2002) Nanocrystal targeting *in vivo*. Proc Natl Acad Sci USA 99:12617-12621

Alivisatos AP (2001) Less is more in medicine. Sci Am 285:66-73

Andersson J, Rosenholm J, Areva S, Linden M (2004) Influences of material characteristics on ibuprofen drug loading and release profiles from ordered micro- and mesoporous silica matrices. Chem Mater 16:4160-4167

Angelos S, Johansson E, Stoddart JF, Zink JI (2007) Mesostructured silica supports for functional materials and molecular machines. Adv Funct Mater 17:2261-2271

Bagshaw SA, Prouzet E, Pinnavaia TJ (1995) Templating of mesoporous molecular sieves by nonionic polyethylene oxide surfactants. Science 269:1242-1244

Barbe C, Bartlett J, Kong LG, Finnie K, Lin HQ, Larkin M, Calleja S, Bush A, Calleja G (2004) Silica particles: a novel drug-delivery system. Adv Mater 16:1959-1966

Beck JS, Vartuli JC, Roth WJ, Leonowicz ME, Kresge CT, Schmitt KD, Chu CTW, Olson DH, Sheppard EW (1992) A new family of mesoporous molecular sieves prepared with liquid crystal templates. J Am Chem Soc 114:10834-10843

Cai W, Gentle IR, Lu GQ, Zhu JJ, Yu A (2008) Mesoporous silica templated biolabels with releasable fluorophores for immunoassays. Anal Chem 80:5401-5406

Calvo EJ, Forzani ES, Otero M (2002) Study of layer-by-layer self-assembled viscoelastic films on thickness-shear mode resonator surfaces. Anal Chem 74:3281-3289

Caruso F (2001) Nanoengineering of particle surfaces. Adv Mater 13:11-22

Caruso F, Shi X, Caruso RA, Susha A (2001) Hollow titania spheres from layered precursor deposition on sacrificial colloidal core particles. Adv Mater 13:740-744

Casasus R, Marcos MD, Martinez-Manez R, Ros-Lis JV, Soto J, Villaescusa LA, Amoros P, Beltran D, Guillem C, Latorre J (2004) Toward the development of ionically controlled nanoscopic molecular gates. J Am Chem Soc 126:8612-8613

Cavallaro G, Pierro P, Palumbo FS, Testa F, Pasqua L, Aiello R (2004) Drug delivery devices based on mesoporous silicate. Drug Delivery 11:41-46

Charnay C, Begu S, Tourne-Peteilh C, Nicole L, Lerner DA, Devoisselle JM (2004) Inclusion of ibuprofen in mesoporous templated silica: drug loading and release property. Eur J Pharm Biopharm 57:533-540

Chattopadhyay PK, Price DA, Harper TF, Betts MR, Yu J, Gostick E, Perfetto SP, Goepfert P, Koup RA, De Rosa SC, Bruchez MP, Roederer M (2006) Quantum dot semiconductor nanocrystals for immunophenotyping by polychromatic flow cytometry. Nat Med 12:972-977

Dai Z, Voigt A, Leporatti S, Donath E, Dhne L, Möhwald H (2001) Layer-by-layer self-assembly of polyelectrolyte and low molecular weight species into capsules. Adv Mater 13:1339-1342

Daniel MC, Astruc D (2004) Gold nanoparticles: assembly, supramolecular chemistry, quantum-size-related properties, and applications toward biology, catalysis, and nanotechnology. Chem Rev 104:293-346

Decher G (1997) Fuzzy nanoassemblies: toward layered polymeric multicomposites. Science 277:1232-1237

Decher G, Hong JD, Schmitt J (1992) Buildup of ultrathin multilayer films by a self-assembly process. 3. consecutively alternating adsorption of anionic and cationic polyelectrolytes on charged surfaces. Thin Solid Films 210/211:831-835

Deere J, Magner E, Wall JG, Hodnett BK (2003) Adsorption and activity of proteins onto mesoporous silica. Catalysis Letters 85:19-23

Ferrari M (2005) Cancer nanotechnology: opportunities and challenges. Nat Rev Cancer 5:161-171

Fu Q, Rama Rao GV, Ward TL, Lu Y, Lopez GP (2007) Thermoresponsive transport through ordered mesoporous silica/pnipaam copolymer membranes and microspheres. Langmuir 23:170-174

Fu Q, Rao GVR, Ista LK, Wu Y, Andrzejewski BP, Sklar LA, Ward TL, Lopez GP (2003) Control of molecular transport through stimuli-responsive ordered mesoporous materials. Adv Mater 15:1262-1266

Ge L, Möhwald H, Li J (2003) Phospholipase A2 hydrolysis of mixed phospholipid vesicles formed on polyelectrolyte hollow capsules. Chem Eur J 9:2589-2594

Gemeinhart RA, Luo D, Saltzman WM (2005) Cellular fate of a modular dna delivery system mediated by silica nanoparticles. Biotech Prog 21:532-537

Giri S, Trewyn BG, Stellmaker MP, Lin VSY (2005) Stimuli-responsive controlled-release delivery system based on mesoporous silica nanorods capped with magnetic nanoparticles13. Angew Chem Int Edit 44:5038-5044

Hata H, Saeki S, Kimura T, Sugahara Y, Kuroda K (1999) Adsorption of taxol into ordered mesoporous silicas with various pore diameters. Chem Mater 11:1110-1119

Heikkila T, Salonen J, Tuura J, Hamdy MS, Mul G, Kumar N, Salmi T, Murzin DY, Laitinen L, Kaukonen AM, Hirvonen J, Lehto VP (2007) Mesoporous silica material tud-1 as a drug delivery system. Int J Pharm 331:133-138

Hernandez R, Tseng HR, Wong JW, Stoddart JF, Zink JI (2004) An operational supramolecular nanovalve. J Am Chem Soc 126:3370-3371

Holm BA, Bergey EJ, De T, Rodman DJ, Kapoor R, Levy L, Friend CS, Prasad PN (2002) Nanotechnology in biomedical applications. Mol Cryst Liq Cryst 374:589-598

Huo Q, Margolese DI, Ciesla U, Demuth DG, Feng P, Gier TE, Sieger P, Firouzi A, Chmelka BF (1994) Organization of organic molecules with inorganic molecular species into nanocomposite biphase arrays. Chem Mater 6:1176-1191

Inagaki S, Fukushima Y, Kuroda K (1993) Synthesis of highly ordered mesoporous materials from a layered polysilicate. J Chem Soc-Chem Commun:680-682

Izquierdo-Barba I, Martinez A, Doadrio AL, Perez-Pariente J, Vallet-Regí M (2005) Release evaluation of drugs from ordered three-dimensional silica structures. Eur J Pharm Sci 26:365-373

Jansen JC, Shan Z, Marchese L, Zhou W, van der Puil N, Maschmeyer T (2001) A new templating method for three-dimensional mesopore networks. Chem Commun: 713-714

Jin S, Ye K (2007) Nanoparticle-mediated drug delivery and gene therapy. Biotechnol Prog 23:32-41

Karen A, Fisher KDHMJT (2003) Comparison of micro- and mesoporous inorganic materials in the uptake and release of the drug model fluorescein and its analogues. Chem-A Eur J 9:5873-5878

Kim J, Kim HS, Lee N, Kim T, Kim H, Yu T, Song IC, Moon WK, Hyeon T (2008) Multifunctional uniform nanoparticles composed of a magnetite nanocrystal core and a mesoporous silica shell for magnetic resonance and fluorescence imaging and for drug delivery 13. Angew Chem Int Edit 47:8438-8441

Kim S, Lim YT, Soltesz EG, De Grand AM, Lee J, Nakayama A, Parker JA, Mihaljevic T, Laurence RG, Dor DM, Cohn LH, Bawendi MG, Frangioni JV (2004) Near-infrared fluorescent type II quantum dots for sentinel lymph node mapping. Nat Biotech 22:93-97

Koktysh DS, Liang X, Yun B, Pastoriza-Santos I, Matts RL, Giersig M, Serra-Rodrguez C, Liz-Marzn LM, Kotov NA (2002) Biomaterials by design: layer-by-layer assembled ion-selective and biocompatible films of TiO_2 nanoshells for neurochemical monitoring. Adv Funct Mater 12:255-265

Kosmulski M (2004) pH-dependent surface charging and points of zero charge II. update. J Colloid Interface Sci 275:214-224

Kresge CT, Leonowicz ME, Roth WJ, Vartuli JC, Beck JS (1992) Ordered mesoporous molecular-sieves synthesized by a liquid-crystal template mechanism. Nature 359: 710-712

Kumar R, Roy I, Ohulchanskyy TY, Goswami LN, Bonoiu AC, Bergey EJ, Tramposch KM, Maitra A, Prasad PN (2008) Covalently dye-linked, surface-controlled, and bioconjugated organically modified silica nanoparticles as targeted probes for optical imaging. ACS Nano 2:449-456

Lai C-Y, Trewyn BG, Jeftinija DM, Jeftinija K, Xu S, Jeftinija S, Lin VSY (2003) A mesoporous silica nanosphere-based carrier system with chemically removable CdS nanoparticle caps for stimuli-responsive controlled release of neurotransmitters and drug molecules. J Am Chem Soc 125:4451-4459

Larson DR, Zipfel WR, Williams RM, Clark SW, Bruchez MP, Wise FW, Webb WW (2003) Water-soluble quantum dots for multiphoton fluorescence imaging *in vivo*. Science 300:1434-1436

Li, Shi, Hua, Chen, Ruan, Yan (2003a) Hollow spheres of mesoporous aluminosilicate with a three-dimensional pore network and extraordinarily high hydrothermal stability. Nano Letters 3:609-612

Li D, Cui Y, Wang K, He Q, Yan X, Li J (2007) Thermosensitive nanostructures comprising gold nanoparticles grafted with block copolymers. Adv Funct Mater 17:3134-3140

Li Y, Shi J, Chen H, Hua Z, Zhang L, Ruan M, Yan J, Yan D (2003b) One-step synthesis of hydrothermally stable cubic mesoporous aluminosilicates with a novel particle structure. Microporous and Mesoporous Materials 60:51-56

Lin YS, Tsai CP, Huang HY, Kuo CT, Hung Y, Huang DM, Chen YC, Mou CY (2005) Well-ordered mesoporous silica nanoparticles as cell markers. Chem Mater 17: 4570-4573

Lin YS, Wu SH, Hung Y, Chou YH, Chang C, Lin ML, Tsai CP, Mou CY (2006) Multi-functional composite nanoparticles: magnetic, luminescent, and mesoporous. Chem Mater 18:5170-5172

Liu N, Chen Z, Dunphy DR, Jiang YB, Assink RA, Brinker CJ (2003a) Photoresponsive

nanocomposite formed by self-assembly of an azobenzene-modified silane13. Angew Chem Int Edit 42:1731-1734
Liu N, Assink RA, Smarsly B, Brinker CJ (2003b) Synthesis and characterization of highly ordered functional mesoporous silica thin films with positively chargeable $-NH_2$ groups. Chem Commun:1146-1147
Lu J, Choi E, Tamanoi F, Zink JI (2008) Light-activated nanoimpeller-controlled drug release in cancer cells. Small 4:421-426
Mal NK, Fujiwara M, Tanaka Y (2003a) Photocontrolled reversible release of guest molecules from coumarin-modified mesoporous silica. Nature 421:350-353
Mal NK, Fujiwara M, Tanaka Y, Taguchi T, Matsukata M (2003b) Photo-switched storage and release of guest molecules in the pore void of coumarin-modified MCM-41. Chem Mater 15:3385-3394
Manzano M, Balas F, Civantos A, Vallet-Regí M (2006) Proceedings of the 20th european conference on biomaterials, Nantes
Michalet X, Pinaud FF, Bentolila LA, Tsay JM, Doose S, Li JJ, Sundaresan G, Wu AM, Gambhir SS, Weiss S (2005) Quantum dots for live cells, *in vivo* imaging, and diagnostics. Science 307:538-544
Nguyen TD, Tseng HR, Celestre PC, Flood AH, Liu Y, Stoddart JF, Zink JI (2005) A reversible molecular valve. Proc Natl Acad Sci USA 102:10029-10034
Pankhurst QA, Connolly J, Jones SK, Dobson J (2003) Applications of magnetic nanoparticles in biomedicine. J Phys D: Appl Phys 36:R167-R181
Parak WJ, Gerion D, Pellegrino T, Zanchet D, Micheel C, Williams SC, Boudreau R, Le Gros MA, Larabell CA, Alivisatos AP (2003) Biological applications of colloidal nanocrystals. Nanotechnology 14:R15-R27
Patel K, Angelos S, Dichtel WR, Coskun A, Yang Y-W, Zink JI, Stoddart JF (2008) Enzyme-responsive snap-top covered silica nanocontainers. J Am Chem Soc 130: 2382-2383
Paul W, Sharma C (2001) Porous Hydroxyapatite nanoparticles for intestinal delivery. Trends in Biomaterials and Artificial Organs 14:37-38
Pelton R (2000) Temperature-sensitive aqueous microgels. Adv Colloid Interfac 85:1-33
Pelton RH, Pelton HM, Morphesis A, Rowell RL (1989) Particle sizes and electrophoretic mobilities of poly (*N*-isopropylacrylamide) latex. Langmuir 5:816-818
Qu F, Zhu G, Huang S, Li S, Sun J, Zhang D, Qiu S (2006) Controlled release of captopril by regulating the pore size and morphology of ordered mesoporous silica. Microporous and Mesoporous Materials 92:1-9
Roy I, Ohulchanskyy TY, Pudavar HE, Bergey EJ, Oseroff AR, Morgan J (2003) Ceramic-based nanoparticles entrapping water-insoluble photosensitizing anticancer drugs: a novel drug carrier system or photodynamic therapy. Journal of American Chemical Society 125:7860-7865
Saha S, Leung KCF, Nguyen TD, Stoddart JF, Zink JI (2007) Nanovalves. Adv Funct Mater 17:685-693
Sakamoto Y, Kim TW, Ryoo R, Terasaki O (2004) Three-dimensional structure of large-pore mesoporous cubic iad silica with complementary pores and its carbon replica by electron crystallography. Angew Chem Int Edit 43:5231-5234
Sierocki P, Maas H, Dragut P, Richardt G, Vogtle F, De Cola L, Brouwer F, Zink JI (2006) Photoisomerization of azobenzene derivatives in nanostructured silica. J Phys Chem B 110:24390-24398

Sing KSW, Everett DH, Haul RAW, Moscou L, Pierotti RA, Rouquerol J, Siemieniewska T (1985) Reporting physisorption data for gas/solid systems with special reference to the determination of surface area and porosity. Pure Appl Chem 57:603-619

Trewyn BG, Slowing II, Giri S, Chen H-T, Lin VSY (2007) Synthesis and functionalization of a mesoporous silica nanoparticle based on the sol–gel process and applications in controlled release. Accounts Chem Res 40:846-853

Trewyn BG, Whitman CM, Lin VSY (2004) Morphological control of room-temperature ionic liquid templated mesoporous silica nanoparticles for controlled release of antibacterial agents. Nano Letters 4:2139-2143

Vallet-Regí M, Rámila A, del Real RP, Pérez-Pariente J (2001) A new property of mcm-41: drug delivery system. Chem Mater 13:308-311

Vallet-Regí M, Balas F, Arcos D (2007) Mesoporous materials for drug delivery. Angew Chem Int Edit 46:7548-7558

Washmon-Kriel L, Jimenez VL, Balkus KJ (2000) Cytochrome C immobilization into mesoporous molecular sieves. Journal of Molecular Catalysis B: Enzymatic 10: 453-469

Wu SH, Lin YS, Hung Y, Chou YH, Hsu YH, Chang C, Mou CY (2008a) Multifunctional mesoporous silica nanoparticles for intracellular labeling and animal magnetic resonance imaging studies. ChemBioChem 9:53-57

Wu T, Zhang Y, Wang X, Liu S (2008b) Fabrication of hybrid silica nanoparticles densely grafted with thermoresponsive poly (*N*-isopropylacrylamide) brushes of controlled thickness via surface-initiated atom transfer radical polymerization. Chem Mater 20:101-109

Yang Q, Wang S, Fan P, Wang L, Di Y, Lin K, Xiao F-S (2005) pH-responsive carrier system based on carboxylic acid modified mesoporous silica and polyelectrolyte for drug delivery. Chem Mater 17:5999-6003

Yang Y, Yan X, Cui Y, He Q, Li D, Wang A, Fei J, Li J (2008) Preparation of polymer-coated mesoporous silica nanoparticles used for cellular imaging by a "graft-from" method. J Mater Chem 18:5731-5737

Yih TC, Al-Fandi M (2006) Engineered nanoparticles as precise drug delivery systems. J Cell Biochem 97:1184-1190

You Y-Z, Kalebaila KK, Brock SL, Oupicky D (2008) Temperature-controlled uptake and release in PNIPAM-modified porous silica nanoparticles. Chem Mater 20:3354-3359

Zeng W, Qian X-F, Zhang Y-B, Yin J, Zhu Z-K (2005) Organic modified mesoporous MCM-41 through solvothermal process as drug delivery system. Mater Res Bull 40: 766-772

Zhang L, Qiao S, Jin Y, Cheng L, Yan Z, Lu GQ (2008) Hydrophobic functional group initiated helical mesostructured silica for controlled drug release. Adv Funct Mater 18:3834-3842

Zhang LZ, Tang GQ, Gao BW, Zhang GL (2004) Spectroscopic studies on the excited-state properties of the light-induced antiviral drug hypocrellin a loaded in the meso-porous solid. Chem Phys Lett 396:102-109

Zhao D, Feng J, Huo Q, Melosh N, Fredrickson GH, Chmelka BF, Stucky GD (1998) Triblock copolymer syntheses of mesoporous silica with periodic 50 to 300. Angstrom Pores Science 279:548-552

Zheng S, Tao C, He Q, Zhu H, Li J (2004) Self-assembly and characterization of polypyrrole and polyallylamine multilayer films and hollow shells. Chem Mater 16: 3677-3681

Zhu S, Zhou Z, Zhang D (2007) Control of drug release through the in situ assembly of stimuli-responsive ordered mesoporous silica with magnetic particles. ChemPhysChem 8:2478-2483

Zhu Y, Shi J, Li Y, Chen H, Shen W, Dong X (2005) Hollow mesoporous spheres with cubic pore network as a potential carrier for drug storage and its *in vitro* release kinetics. J Mater Res 20:54-61

Zou H, Wu S, Shen J (2008) Polymer/silica nanocomposites: preparation, characterization, properties, and applications. Chem Rev 108:3893-3957

2

Nanostructured Functional Inorganic Materials Templated by Natural Substances

Yuanqing Gu, Jianguo Huang*

Department of Chemistry, Zhejiang University, Hangzhou, Zhejiang, 310027, China.
*E-mail: jghuang@zju.edu.cn

Naturally-produced sophisticated hierarchal structures and the astonishing properties of biological substances are difficult to obtain artificially, even with the most technologically advanced synthetic methodologies. As the needs for the development of advanced materials with improved performance characteristics become increasingly important, the potential of natural substances for material design and fabrication is being actively explored. The combination of versatile synthetic chemical strategies and biological assemblies provides precise replication of natural substances with inorganic/organic matrices, resulting in man-made materials which memorize the initial biological structures. Natural substances provide tailored template structures to be transcribed by guest matrices, or direct the organization of guest substrates in a certain order. This chapter will give a brief review of the research done by means of biotemplate synthesis, which is an emerging, unique approach for the synthesis and organization of inorganic/organic micro- and nano-scale building blocks into well-defined architectures and hence functional bio-inspired materials.

2.1 Introduction

Fabrication of advanced functional nanomaterials with various application potentials, such as extremely sensitive sensor devices to probe confined environments and multiplexed techniques for high-throughput analysis, is one of the most prominent technological needs today. The ultimate realization of such advanced materials is based on methodologies to synthesize and organize the related building blocks into controlled geometries on the micro- or even nano-scale. During the past few years, advanced functional materials with well-defined nanostructures were synthesized, showing unique optoelectronic, magnetic, or catalytic properties (Sotiropoulou et al., 2008; van Bommel et al., 2003). Rather than constructing nanostructured architectures with guest building blocks, transcription of the structure from a variety of templates to the guest materials provides an easy pathway to retain otherwise unattainable structures. One can envision the use of template-synthesized hollow spheres for the controlled release of a series of substances, as well as their use as light, but strong filler material or as microreactors. Also, template-derived chiral materials can be applied in chiral catalysis, and fine fibrous materials possess potential in the fabrication of nanotechnological devices. In particular, biological substances distinguish from the vast numbers of templates.

Nature is a rich and constant source providing amazingly versatile living organisms, which are formed by self-assembly of highly ordered units and possess inherently sophisticated and hierarchical micro-to-nano-scale features. Through transcribing natural biological substances, advanced materials can inherit the unique complex multilevel micro- or/and nano-structures and morphologies as well as the excellent properties of the natural substances. Hence, replication using natural substances as templates, the so-called biotemplate synthesis approach, presents an easy, low-cost, and environmentally friendly pathway for tailoring and preparing advanced functional man-made materials (Sanchez et al., 2005; Pouget et al., 2007).

Within the last ten years, there had been incredible progress in the field of biotemplate synthesis, which sought to either replicate the morphological characteristics and functionalities of biological substances, or to use a biological structure with specific physicochemical and/or morphological attributes to guide the assembly of guest materials. Typically, the process of replication is divided into several steps. Firstly, a natural template substrate, which consists of either preformed or self-assembled entities, is brought into contact with a precursor or small particles of the guest materials. Notably, this step can occur in solution as well as in the gas phase. Then, the guest materials are deposited on the inner or outer surface of the template. Finally, the organic template material may then be removed by, for example, heat treatment (Kawahashi, Matijević, 1991), microwave irradiation (Gallis, Landry, 2001), or washing with organic solvents (Lu et al., 2001), resulting in the isolation of the guest material with a morphology directly related to the template. Heat treatment, which is generally used for the removal of the organic template, can simultaneously lead to calcination of the inorganic material, giving a harder and more stable inorganic product. The transcription process typically leads to the generation of either a negative, positive (or hollow), or exact copy of the template substrate. Naturally, template removal is optional; the template can either be retained to act as a scaffold or provide carbon resources through pyrolysis, and the preceding organic-inorganic hybrid material may prove to be even more interesting from the application point of view.

2.2 Metal Oxide Nanomaterials

Various guest materials such as metal oxides, metal nanoparticles, and even organics are templated by a large variety of biological species to produce nanostructured materials. Indeed, researchers have managed to achieve astonishing levels of control over the biological-inorganic/organic interface leading to highly uniform nanostructures, demonstrating that biotemplate synthesis is emerging as an effective new route to organize nanomaterials into one-, two- or three-dimensional technological platforms. Such materials are potentially interesting for sensing applications like photonic devices and biosensors, or as three-dimensional architectures that can be further used for the fabrication of higher-ordered structures. Here, we overview them according to the material type of the resultant products.

2.2.1 Silica Nanomaterials

Silica, also known as silicon dioxide, is most commonly found in nature as sand or quartz, as well as in the cell walls of diatoms. It is the most abundant mineral in the earth's crust. It is a principal component of most types of glass and substances such as concrete. Apart from its hardness, since antiquity, silica has possessed several attractive properties such as inexpensiveness, bio-compatibility, and thermal stability, which have made it widely used as a raw material in modern industry as well as in daily life (Norris, 2007).

Initially in biotemplate synthesis, it was considered that the presence of positive charges in the template substrates facilitated the transcription process, and hence natural proteins with cationic charges were employed (Ono et al., 1998). Collagen, which becomes positively charged at pH lower than 8, well meets this essential element. Meanwhile, it can form a unique fibrous superstructure, where there exists a periodic groove every 67 nm of the fiber. Such interesting natural architecture is copied by adding tetraethylorthosilicate (TEOS) to a buffered saline solution containing collagen fibers, forming a silica replica with a gnarled structure (Fig.2.1) (Ono et al., 1999). In electron microscopic images, the copied gnarled structure with

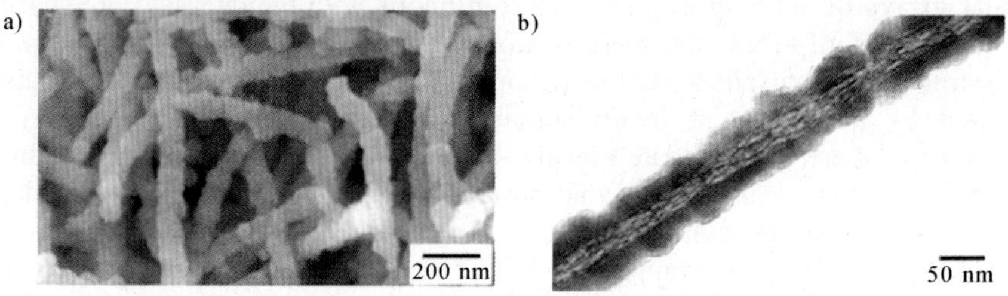

Fig.2.1. a) Image of scanning electron microscopy (SEM) of silica fibers transcribing collagen; b) image of transmission electron microscopy (TEM) of an individual fiber showing that the inner tube contains several individual channels. Reprinted from Ono et al., 1999. Copyright (1999), with permission from the Chemical Society of Japan

a period of 60~80 nm is clearly seen. The silica fibers possess an inner tubular structure which consists of a bundle of individual channels. This result corresponds to the individual fibers which make up the collagen bundles, and thus demonstrates that accurate transcription is achieved. After calcination, organic components are removed and these individual channels remain in the resulting pure inorganic material.

There is considerable interest focused on the production of inorganic materials containing frameworks with well-defined pore networks. Generally, strategies for fabricating such materials mainly depend on the use of templates, the size and nature of which dictate the pore dimensions and morphologies, where natural substances can exhibit their structural advantages. Hierarchically structured zeolite fibers containing ordered pores at the nano- and micro-scopic length scale were successfully synthesized with bacterial templates (Zhang et al., 2000). Here, zeolite nanoparticles were used as building blocks and infiltrated into the ordered void spaces of a bacterial supercellular thread through reverse swelling. Since silica is bio-compatible, so the bacterial template was not affected and the infiltration process was enabled. Silicalite nanoparticles aggregated specifically within the organized micro-architecture, resulting in the construction of a macroporous inorganic framework. The bacterial template was decomposed by subsequent thermal treatment, giving an intact zeolite fiber with ordered macroporous channels lined by 100-nm wide walls of coalesced silicalite nanoparticles.

Diatoms are eukaryotic, unicellular photosynthetic algae with sizes in the 1~100 μm range that can be found in almost any aquatic habitat on earth, and dominate phytoplankton populations and algal blooms in the oceans. Apart from their ecological significance, diatoms are well known for the intricate geometries and spectacular patterns of their silica-based cell walls. These silica cell walls have unique nanostructured patterns which can be hexagonal, rod-shaped, or circular depending on the species (Fig.2.2), and so provide a powerful pathway for the production of nanostructured silica materials (Sumper, Brunner, 2006). Indeed, these intricate porous nanostructures in the silica shell itself have great advantages in zeolitization. By coating the diatoms with zeolite nanoparticles with subsequent attachment and hydrothermal growth of zeolite crystals, diatomaceous earth was zeolitized (Anderson et al., 2000). The initial diatoms consist of shell forms where regular arrays of submicron pores are continuous with the internal pore space of the hollow diatom tubes, and there is around 0.5 μm of silica between the pores. MFI structural typed silica zeolite nanoparticles were dispersed on the diatoms followed by a subsequent hydrothermal growth step within 24 h. Thus, the nanoparticles were evenly coated on the diatom shell surface, retaining the initial morphologies and clear internal void space. In this way, the surface of a diatom structure is successfully zeolitized.

Wood is a natural composite substance mainly composed of cellulose, hemicellulose, and lignin. The tracheidal cells or vessels are distributed over the cross-section of the wood, forming a uniaxial pore channel system which makes wood a perfect candidate for biotemplate synthsis. With this amazing template, Shin et al. prepared hierarchically-ordered amorphous silica ceramics (Shin et al., 2001).

a)
b)

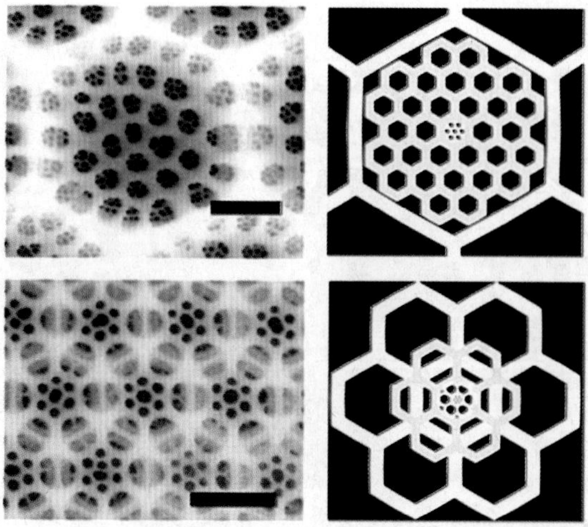

Fig.2.2. a) SEM image of *Coscinodiscus asteromphalus* and the interpretation of the valve pattern by superimposition of three hexagonal silica frameworks. b) SEM image of *Coscinodiscus granii* and the interpretation of the valve pattern by superimposition of four hexagonal silica frameworks (scale bars = 1 μm). Reprinted from Sumper, Brunner, 2006. Copyright (2006), with permission from Wiley

For faithful replication, surfactant was used to support mineralization during the sol–gel process. Here, the modified sol–gel process was applied because the hydrolysis rate can be adjusted easily by changing the solvent ratio and the acidity of the solution to avoid bulk precipitation or gelation of the silicate species during the entire process. After the wood tissue was immersed in a solution containing TEOS and surfactant, the silicate species penetrated the cell wall structures, and then hydrolyzed and condensed around the cellular tissues. Hence, the surfactant micellar structures were incorporated in the silica network, producing organized nanoporous channels (Fig.2.3) by subsequent calcination. The formation of these continuous nanochannels was critical in maintaining the structural integrity, as they provided pathways for the gaseous decomposed organic components to leave without destroying the intact structure. On the contrary, the transcription of the wood structure without the surfactant just resulted in collapsed inorganic material which failed to inherit the initial whole structure after calcination.

Furthermore, with a seed growth strategy taking advantages of the hierarchical porous structure, as illustrated in Scheme 2.1, a self-standing zeolitic tissue was prepared (Dong et al., 2002). Silicalite-1, a pure silica zeolite, was used because the unique microporous framework of such materials may extend the application range, and more importantly, the formation process of zeolitic tissue can mimic the natural petrification of wood more vividly. As a result, the product zeolitic tissue faithfully preserved the initial cellular structure of wood at various hierarchical levels. Moreover, the as-grown silicalite-1 crystals formed mesopores which originated from the intercrystalline voids between the individual zeolite particles or the pinholes in the intergrown zeolite layer.

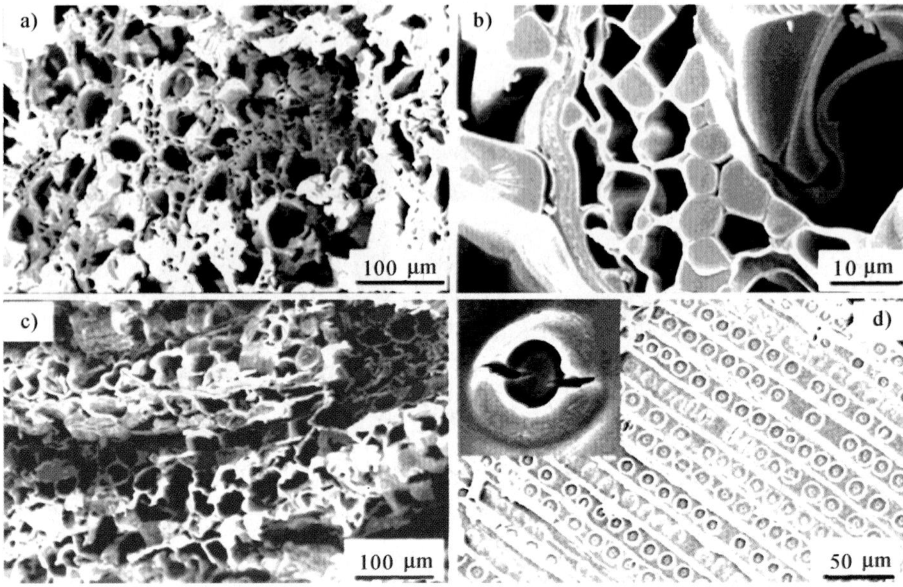

Fig.2.3. SEM images of surfactant-assisted wood samples after calcination. a) Cross-section of poplar at low magnification. b) Cross-section of poplar at higher magnification. c) Cross-section of pine. d) Longitudinal direction of pine. The inset shows the pit structure. Reprinted from Shin et al., 2001. Copyright (2001), with permission from Wiley

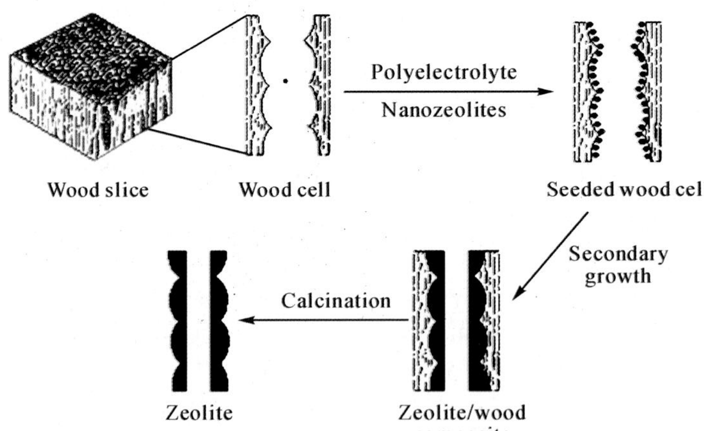

Scheme 2.1. Schematic illustration of the fabrication process of wood tissue-templated hierarchically organized zeolite materials. Reprinted from Dong et al., 2002. Copyright (2002), with permission from Wiley

Biotemplated silica materials are applied not only in synthesizing zeolites. The mild conditions of the sol–gel process and biocompatibility of silica provide a way to immobilize microorganisms while retaining their bio-functionality in silica matrixes. Monodispersed yeast cells are immobilized by a dip-coating process, forming patterned hexagonal arrays of living cells in sol–gel silica films (Fig.2.4). Such ordered two-dimensional arrays of immobilized living yeast cells can screen genomic libraries. The addressability with a resolution of microns allows specific cells containing the gene product of interest to be identified and manipulated. With this system, it is possible to screen libraries based on color or fluorescence.

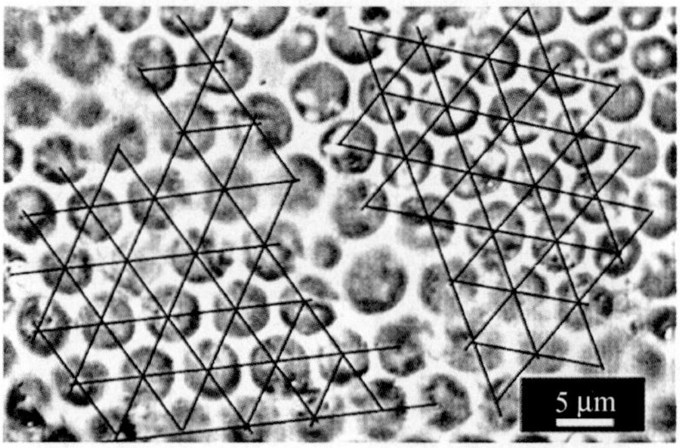

Fig.2.4. Photograph of a silicate film containing a monolayer of hexagonal close-packed yeast cells. The ordered regions are depicted by lines; defects caused by the presence of smaller cells are found in the center and upper left of the image. Reprinted from Chia et al., 2000. Copyright (2000), with permission from the American Chemical Society

Tobacco mosaic virus (TMV) is composed of about 2,130 identical proteins which are helically arranged around a single-strand RNA, forming a 300 nm long tubular structure with an outer diameter of 18 nm and a hollow channel with a diameter of 4 nm (Henrick, Thormton, 1998; Namba et al., 1989; Pattanavek, Stubbs, 1992; Culver et al., 1995; Zaitlin, 2000). Interestingly, when the viruses are protonated or deprotonated by changing the pH of the medium or other conditions, they can attract each other, either head-to-tail, forming a linear aggregation of viruses; or side-by-side, leading to two-dimensional or three-dimensional structures (Knez et al., 2004a; 2004b). TMV possesses a highly defined nanostructure and extraordinary stability compared to other biospecies, making it a very useful template for nanomaterial research. Not only can it withstand temperatures of up to 353 K without destruction of its integral architecture, but also it can be handled within a wide pH range of around 2.8~8.0 for fairly long periods of time. Once it is dried, a crystal-like structure is constructed without destroying the single virions, all of which make TMV a favorable substrate for biotemplate synthesis. Indeed, the addition of a solution of TMV to an acidic ethanolic solution of TEOS gave rise to the formation of a silica layer of about 3 nm around the TMV tubes. Many of the TMV assemblies had self-assembled in an end-to-end structure, giving rise to very long (>1 mm) silica-TMV tubes (Fig.2.5). After calcination, the pure inorganic hollow products were obtained with the initial morphology. In fact, the silica-coated virions exhibited an even higher level of ordered structure by self-assembling into linear chains. In later genetic engineering of the virus, deposition can selectively take place either at predefined positions on the outer surface or even inside the hollow channel.

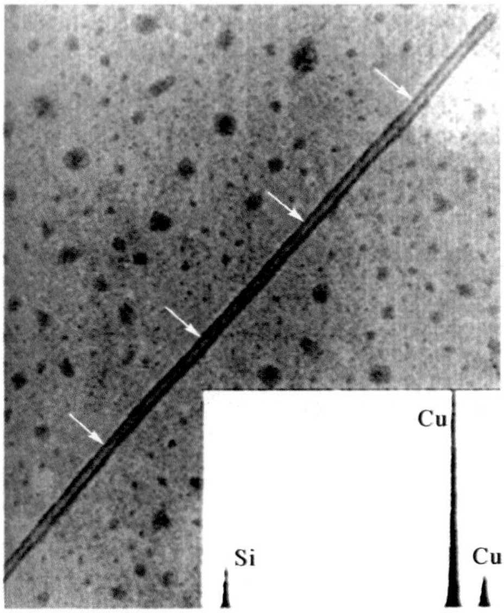

Fig.2.5. TEM image of a tubular-structured silica-coated TMV. The arrows mark the ends of five individual TMV particles, each 300 nm in length. The inset is an energy dispersive X-ray (EDX) spectrum showing the Si peak (the Cu peaks are from the TEM grids). Reprinted from van Bommel et al., 2003. Copyright (2003), with permission from Wiley

In later work, mesostructured silica was fabricated using the tendency of TMV to form nematic liquid crystals at high concentrations (Fowler et al., 2001). With controlled hydrolysis of a mixture of a liquid crystalline gel of TMV particles and the silica precursors TEOS and aminopropyltriethoxysilane (APTES), a thin mineralized gel was formed on the template surface, giving hexagonally ordered TMV fibers in a silica matrix. When the amount of TEOS and APTES used in the transcription was reduced, large numbers of spherical silica nanoparticles of peculiar morphology resulted instead. Particles with diameters of 100~150 nm consisted of an array of 50 nm long, rod-shaped TMV fragments arranged radially around a silica core approximately 35 nm in diameter (Fig.2.6). Many of the nanoparticles were attached to a single silica tube originating from the more complete replication of one of the TMV entities protruding from the silica core. After calcination, nanoparticles of similar size containing radially arranged channels were obtained. Although the origin for the formation of the spherical particles is assumed to lie in the disruption of the liquid-crystalline phase, the exact mechanism is not yet fully understood. In other studies, the filamentous fd virus has also been used as biological templates for some interesting silica structures (Zhang, Buitenhuis, 2007). By hydrolyzing TEOS under acidic conditions, silica nanorods, silica nanowires, and bow-tie-shaped silica bundles are obtained through electrostatic and/or hydrogen-bonding interactions between silica precursor molecules and the organic substrate.

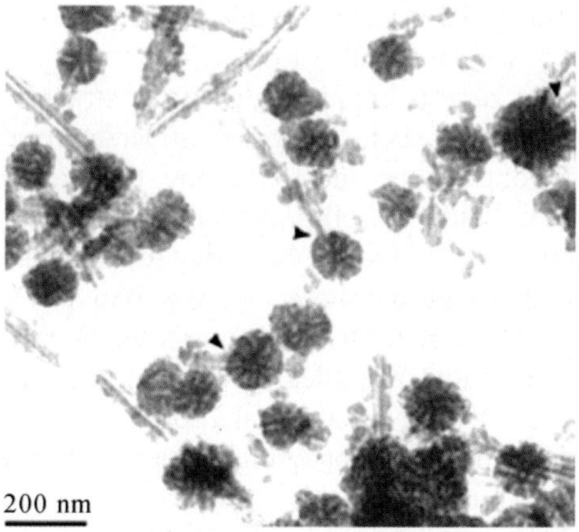

Fig.2.6. TEM image of silica-TMV nanoparticles with radially patterned interiors, together with threadlike mesostructures. The arrows denote nanoparticles with surface-attached threads. Reprinted from Fowler et al., 2001. Copyright (2001), with permission from Wiley

Instead of cells, Mann et al. focused on cuttlebone, obtained from the commonly known cuttlefish *Sepia officinalis*, as a template substrate for novel inorganic material syntheses. The cuttlebone consists of aragonite ($CaCO_3$) and β-chitin, possessing macroporous structures. Demineralization leaves only the β-chitin framework, which can be remineralized by immersing it in a sodium silicate solution with several subsequent immersions in ethanol/water and ethanol/ammonium hydroxide mixtures. After calcination, the pure inorganic material which faithfully memorizes the chamber-like architecture of the initial cuttlebone is isolated (Ogasawara et al., 2000).

As a different kind of substrate, native pollen grains, a ubiquitous and inexpensive material, have average length of *ca.* 25 μm and exhibit a species-specific ellipsoidal morphology consisting of four longitudinal segments with a foam-like surface structure. The tough outer shell (exine) of pollen grains is replicated without consequent loss of the fine structure by both amorphous silica and crystalline particles, such as calcium carbonate and calcium phosphate minerals, resulting in high-fidelity hollow inorganic replicas which demonstrate applicability as a potential drug delivery system with controlled release properties as well as a possibility for post-synthetic functionalization with magnetic or metallic nanoparticles (Hall et al., 2003). The inorganic replication here was achieved by soaking freeze-dried mixed-flower pollen in various metastable solutions followed by thermal removal of the biological template to produce silica, calcium phosphate, or calcium carbonate facsimiles. The resulting morphologically complex colloids exhibit high surface area and mesoporosity that arise as a consequence of the complex foam-like surface morphology of the native pollen grains and outgassing of organic components during thermal degradation. Such unique products not only can be functionalized by impregnation of nanoparticles such as silver or magnetite to respectively produce metallic or magnetic derivatives, but also are useful in controlled release applications due to the silica, calcium carbonate, and calcium phosphate, which are known potentially bioactive materials.

Biomolecules present various unique higher-order structures, so they should act as ideal templates to create such inorganic superstructures. Among versatile biomolecules, DNA is one of the most amazing template substrates due to its double-stranded structure with a well-regulated micrometer length and uniform 2-nm diameter which cannot easily be prepared by artificial polymers or low-molecular-weight gels. Furthermore, in the native DNA system, several topologically different higher-order DNA structures can be found. Such an attractive natural template was replicated by silica, giving unique structures reflecting the higher-order structures of DNA (Numata et al., 2004c). To use DNA as template, a serious problem arises from the mismatching properties between DNA and TEOS (the precursor of silica). To overcome this, molecule 2 illustrated in Scheme 2.2 was designed. This molecule bears one ammonium group and one guanidinium group, with the expectation that intermolecular formation of an ion pair with the guanidinium group would proceed in preference to an intramolecular one with the ammonium group. By modification of molecule 2, DNA can perform as a new "cationic" template for sol–gel transcription, and the resultant silica materials precisely transcribe the high-order structures of DNA. The obtained silica structures are changed drastically, reflecting the conformationally transformed template structures, including partial coiled-coil, coiled-coil, and circular forms.

Schizophyllan (SPG), a natural polysaccharide produced by the fungus *Schizophyllum commune*, has a repeating unit consisting of three β-(1-3) glucoses and one β-(1-6) glucose side-chain linked at every third main-chain glucose. SPG adopts a triple helix (t-SPG) in nature. By dissolving in dimethylsulfoxide (DMSO), this triple helix can be dissociated into a single chain (s-SPG), and *vice versa* by exchanging DMSO for water. During the renaturating process from s-SPG to t-SPG, hydrophobic polymers or nanoparticles can be entrapped in the cavity with the aid of hydrophobic force, giving novel nanocomposites having unique one-dimensional architectures (Numata et al., 2005a; 2004a; 2004b; Bae et al., 2005; Li et al., 2005; Hasegawa et al., 2005). Similarly, hydrophobic TEOS is also entrapped in the SPG one-dimensional cavity, followed by sol–gel polycondensation taking place inside the cavity, resulting in a silica nanofiber with uniform diameter which shows water-solubility as well as biocompatibility for biological applications (Numata et al., 2005b). The side-chain glucose group of SPG can be chemically-functionalized, and the resultant silica nanofiber can be easily manipulated by a supramolecular strategy (Koumoto et al., 2002).

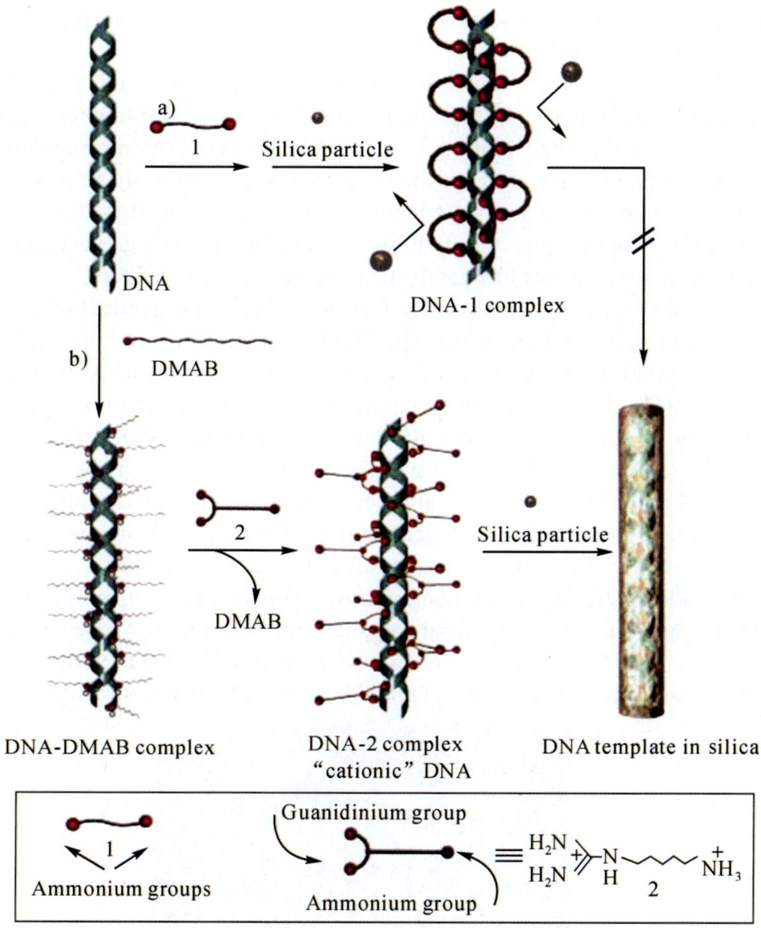

Scheme 2.2. Schematic representation of the DNA surface transformation. Reprinted from Numata et al., 2004c. Copyright (2004), with permission from Wiley

2.2.2 Titania Nanomaterials

Titania, also called as titanium dioxide, is one of the most technologically versatile of oxides, exhibiting various attractive optical, chemical, electrical, thermal, and biological properties. Biotemplate synthesis brings such fascinating characteristics with uniquely intricate functional three-dimensional morphologies and structures of natural substances, giving advanced titania materials with wide ranges of applications. The cellular structures of wood were transformed into titania ceramics by repeated infiltration of titanium tetra (iso-propoxide) into the wood tissue and subsequent hydrolysis (Mizutani et al., 2005). Similarly, a green leaf-derived titania photocatalyst was fabricated, achieving an extremely high light harvesting efficiency due to the inherited structural features favorable for light harvesting from macro- to nano-scales (Li et al., 2009).

As an alternative process, titania hollow fibers with porous walls were prepared by a chemical solution deposition technique using TiF_4 aqueous solutions with textiles and wools (Imai et al., 2000; Mahltig et al., 2005). The deposition process occurred through heterogeneous nucleation on the cotton surface, and the shape of the anatase fibers reflected that of the initial structure. The walls of the resulting fibers are mesoscopically porous, thus the tailored structures possess great potential for various practical applications like catalysts, photocatalysts, and filters.

Avian eggshells are known for being formed by layered organization of calcified shell and organic eggshell membranes (ESMs) containing collagen types I, V, and X, and glycosaminoglycans (Dennis et al., 2000). It is noteworthy that eggshell membranes (ESMs) that consist of the outer shell membrane and inner shell membrane are stable in aqueous and alcoholic media and undergo pyrolysis upon heating (Hincke et al., 2000). The outer shell membrane, which can be easily isolated from eggshells, was employed as a removable template, and interwoven shell membrane titania fibers were fabricated via sol–gel coating followed by heat treatment (Yang et al., 2002; Dong et al., 2007b). The resulting titania network is composed of interwoven and coalescing titania fibers whose diameters range from 0.3 to 1.2 μm (Fig.2.7). Compared with the initial eggshell membrane, the pore sizes of the macropores are significantly decreased, which is considered due to shrinkage caused by template removal. Actually, the titania fibers are of tubular structure (Fig.2.7c).

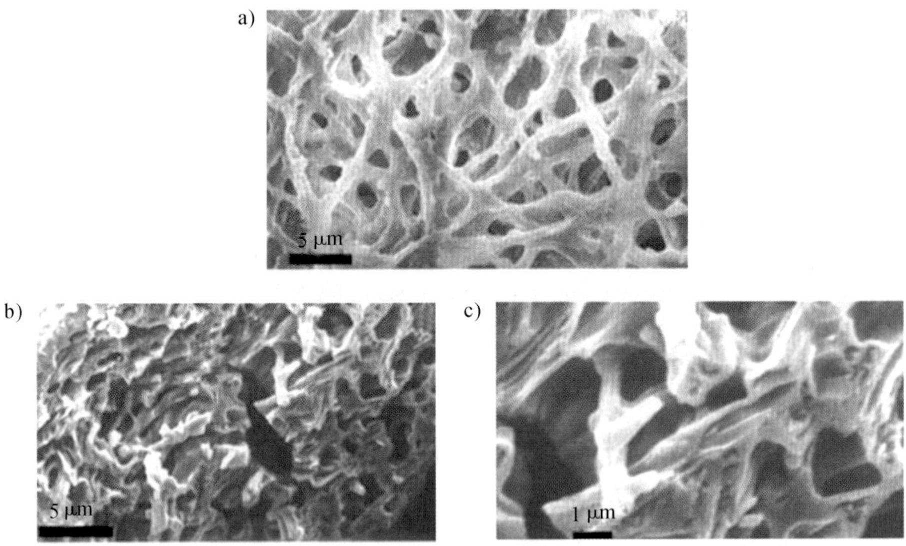

Fig.2.7. SEM images of the titania network derived from the initial eggshell membrane: a) overview of the surface; b) overview of the cross section; c) higher magnification of the cross-section showing the broken hollow tubes. Reprinted from Yang et al., 2002. Copyright (2002), with permission from Wiley

These results indicate the formation of bio-inspired hierarchically ordered titania materials, which will have interesting applications in certain fields such as photocatalysis, gas sensors, antistatic coating, and dye-sensitized solar cells.

For synthesizing thinner anatase nanowires consisting of mesoporous titania networks, unique bacterial cellulose (BC) is used as the natural biological template (Zhang, Qi, 2005). BC is identical to plant cellulose in many aspects. Apart from the different molecular structures, BC membrane has an ultrafine network structure comprising interwoven nanofibers with diameters less than 100 nm (Figs.2.8a and 2.8b), which is much smaller than the diameter of typical plant cellulose bundles (~10 μm), and exhibits unique properties, including high crystallinity, high water holding capacity, high tensile strength, and mouldability during formation. With the surface sol–gel process, the BC-titania hybrid membranes obtained clearly exhibit interconnected ribbon-like nanofibers that faithfully replicate the structure and morphology of the original BC template. Meanwhile, the conductivity is improved, leading to a high-magnification image with higher quality (Fig.2.8c). After calcination, the initially amorphous titania was turned into anatase, but the original ribbon-like morphology failed to be preserved, presumably due to the inability of the very thin the ribbons to withstand the considerable shrinkage occurring during template removal. It is noted that the traditional macroporous titania networks, such as the ESM templated ones mentioned above, showed both crystallinity and surface areas

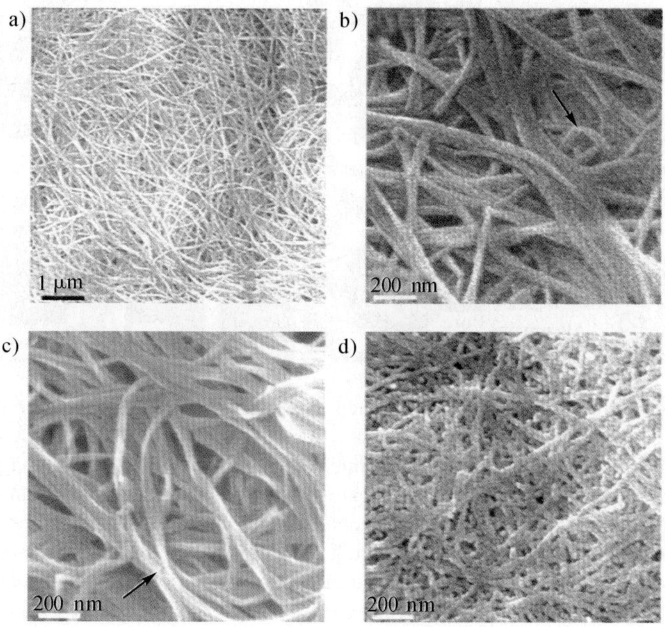

Fig.2.8. a), b) SEM images of BC membranes; c) BC-titania hybrid membranes; d) titania networks templated by BC membranes. Arrows indicate the twisted structure of ribbon-like nanofibers. Reprinted from Zhang, Qi, 2005. Copyright (2005), with permission from the Royal Society of Chemistry

comparable to the present mesoporous titania networks but much larger diameters of the interwoven fibers (typically ~1 mm). Therefore, thanks to the larger accessible surface areas, enhanced photocatalytic activity is realized. The novel titania nanowire networks may find potential applications in areas including photocatalysis, photovoltaics, and bone-tissue engineering.

As an alternative approach, atomic layer deposition (ALD) can also be applied to form a thin film coating on some biological species, such as TMV and ferritin (Knez et al., 2006). To coat such biospecies with thin films, the template substances are dried and transferred into the ALD chamber for film deposition. By repeating the adsorption/hydrolysis cycle with corresponding gaseous chemicals, ultrathin films are formed layer-by-layer on template surface. Various angles of the resultant structure can be clearly seen (Fig.2.9). In the cross-section view given here, a disk of destroyed TMV (circular particle) embedded in an amorphous titania film can be seen. The titania covering the interior channel appears to be hollow with a pore diameter of 1.0~1.5 nm and a wall thickness of 1 nm, indicating that the tubular structure of the initial TMV template is carefully retained. Thus, metal oxide nanotubular material is prepared with a natural template via ALD deposition. Compared with carbon nanotubes, the obtained products possess even finer structure with good stability due to the biomolecular shell retained inside, which presents great application potential.

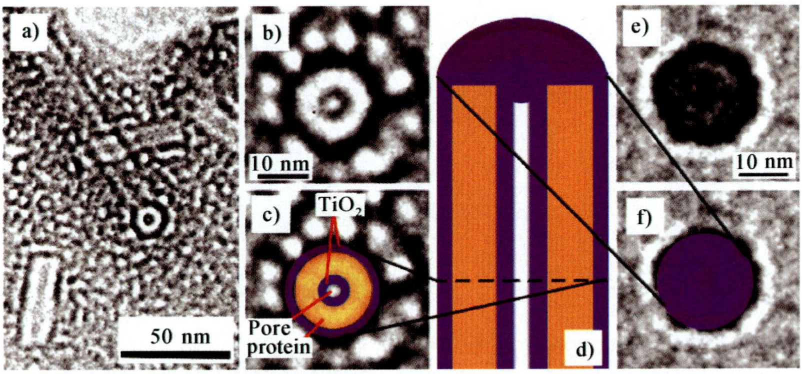

Fig.2.9. a) TEM image of TMV coated by titania using ALD without successive ultrasonication. b) Magnification of a further titania-covered TMV disk, showing a hollow area inside the titania-covered interior channel of the virus. c) Recolored image of b). The orange circle represents the viral protein sheath and the blue color shows the titania coating of the viral surface (outer surface and channel surface). The surrounding gray area is the embedding amorphous titania film. d) Sketch of a cross-section of a titania-covered TMV with the same colors as in image c). e) Magnification of a further titania-covered TMV disk showing a clogged interior channel of the virus. f) Recolored image of e). Reprinted from Knez et al., 2006. Copyright (2006), with permission from the American Chemical Society

Different from just coating a film on the natural substance template surface, Raymond et al. provide a synthetic chemical conversion process, involving a shape-preserving metathetic gas/solid reaction, to overcome the compositional limitations of natural bioclastic structures (Unocic et al., 2004). The silica-based frustules of *Aulacoseira* diatoms were used as the bioclastic structural template and were turned into the technologically important oxide titania. The silica shell walls of diatoms were transformed into titania without destroying the original structures and morphologies, according to the following net metathetic reaction:

$$TiF_4\,(g) + SiO_2\,(s) \rightarrow TiO_2\,(s) + SiF_4\,(g) \qquad (2.1)$$

Here, solid TiF_4 was utilized as a low-temperature source of TiF_4 vapor.

As the representative SEM images of the corresponding samples show (Fig.2.10), the morphologies and structures of the starting *Aulacoseira* diatom frustules are retained with high quality. Details such as the cylindrical shape, rows of fine pores, and narrow V-shaped channels decorating the side walls, as well as the circular hole with a protruding outer rim exhibited on the end faces, can clearly be seen in the resultant product. This work demonstrates that a metathetic halide gas/solid reaction may be used to convert a biologically self-assembled three-dimensional structure into a new nanocrystalline material without loss of the bioclastic shape or fine features. With alternative thermodynamically-favored metathetic reactions to generate three-dimensional assemblies with a wide variety of shapes and functional chemistries, this approach can also be applied to other bioclastic or biomimetic preforms. For example, this titania conversion process can be applied to silica-based materials, including colloidal crystals, membranes, or structures obtained by silicon micromachining.

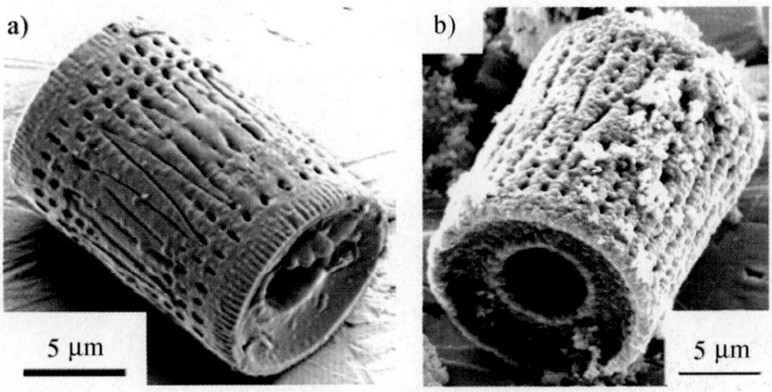

Fig.2.10. SEM images of *Aulacoseira* diatom frustules: a) before treatment; b) after exposure to TiF_4 (g) for 2 h at 623 K. Reprinted from Unocic et al., 2004. Copyright (2004), with permission from the Royal Society of Chemistry

Since some natural biological substances appear to be delicate and are easily affected or even destroyed under "hard" conditions, Wang et al. proposed a supercritical fluids (SCFs)-aided method for biotemplate synthesis (Wang et al., 2005b). SCFs exhibit several attractive features such as low viscosity, high diffusivity, zero surface tension, and tunable solvent power, and have been demonstrated to be excellent solvents and carriers to distribute and/or impregnate precursors into porous substrates for preparation of composites and/or template synthesis of structural materials (Wakayama et al., 2001; Crowley et al., 2003; Morley et al., 2002; Cooper, 2003). Traditional replication strategies such as the wet chemistry route failed to faithfully copy their microscopic structures because irregular aggregates of particles coat the biotemplate surfaces in an uneven way and severely mask the fine surface structure of the template. The special properties of SCFs can help the precursor molecules enter every fine void of the template and react with the surface-active groups, resulting in faithfully replicated surface structures with uniform, compact, and smooth films. Every corner of pollen grains is precisely replicated by this methodology (Fig.2.11). This replication process provides a promising route not only for the design and fabrication of complex structured materials with natural substances, but also for the coating or deposition of delicate biological organizations and cultural relics for the purpose of fixing or conservation. This SCF replication process can also be extended to the synthesis of other inorganic materials such as silica.

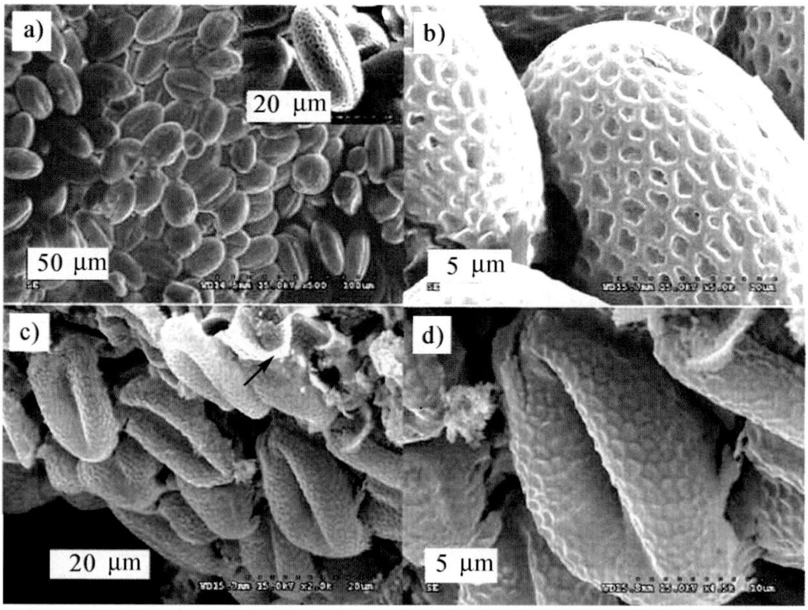

Fig.2.11. SEM images: a) cole pollen grains (inset is the enlarged image); b) coated pollen grains; c), d) titania replicas. Reprinted from Wang et al., 2005b. Copyright (2005), with permission from the Royal Society of Chemistry

With sol–gel transcription processes, various advanced materials were successfully synthesized using biological templates. Unfortunately, these processes usually occur in organic solvents such as dichloromethane, ethanol, and butanol in the presence of only a small amount of water, which limits further application of some organic templates from lipids and biomolecules that can self-assemble in water.

To break through this limitation, lipid was iced to perform as template substrates, giving well-defined transition metal oxide nanotubes achieved in water (Ji, Shimizu, 2005). In the typical approach, the aqueous dispersion of lipids was first frozen to an iced solid state in liquid nitrogen, followed by addition of metal oxide precursor solution at 253 K. Then, the whole reaction system was maintained at 273 K for 2 weeks. This aqueous sol–gel transcription has great significance in the fabrication of inorganic materials from biotemplates, especially in aqueous dispersions.

2.2.3 Tin Oxide Nanomaterials

Marder et al. reported the fabrication of three-dimensional nanostructured tin oxide with *Aulacoseira* diatoms as scaffold using an automatic surface sol–gel like process (Weatherspoon et al., 2007). Thin coatings of nanocrystalline tin oxide (SnO_2) 50 nm thick were applied to the cylindrical, nanostructured silica valves of the biosilica substance by dendritic amplification of surface hydroxyl groups first, and then layer-by-layer deposition of a tin alkoxide was followed by an automated surface sol–gel process. Finally, the specimens were fired in air at 973 K for 2 h and the product shown in Fig.2.12 was obtained. Selected-area electron diffraction (SAED) analysis of the coating exhibited a clear pattern in agreement with cassiterite SnO_2 (Fig.2.12b). And the SnO_2 coating consisted of (7.5±1.5) nm diameter crystallites. Taking advantage of the SnO_2 component and obtaining an interesting structure, the resultant product was fabricated into a gas sensor (Fig.2.12a), and the related NO-sensing capability was tested by exposure to gas mixtures with 3×10^{-6}, 5×10^{-6}, or 8×10^{-6} NO at 623 K. Compared with those reported for other nanocrystalline SnO_2 sensors (Hyodo et al., 2005), the response (rise) and recovery (decay) times of the current SnO_2 sensor (defined as the times needed to reach 90% of the total signal change) were respectively measured as 12 s and 32 s, which are comparable or even faster. This developed protocol is considered to possess general applicability to complex biosilicas or various synthetic silica template substrates.

In a different approach, nanocrystalline SnO_2 can also be fabricated as a hierarchical material with ESM due to the glycoprotein mantle covering the template surface (Dong et al., 2007d). This mantle plays a key role in the tiomorphic SnO_2 hierarchy because it is rich in polyanions exhibiting a variety of keratan sulfate epitopes and dermatan sulfate (Yang et al., 2002; Fernandez et al., 1997; Wong et al., 1984). The amorphous Sn colloid precursors adhere to the template ESM through the soaking procedure. The quantities of carboxylic, hydroxyl, and amino residues contained in the glycoprotein mantle of ESM fibers promote the adsorption of stannic cations from the colloid medium by strong affinity involving multiple hydrogen bonds, van der Waals interactions, resulting in Sn-ESM hybrids. Subsequently, the sample is calcined to remove the template substrates, and the as-deposited amorphous tin (IV) oxo species ripen into well-defined SnO_2 nanocrystallites, congregating connective tubular structures and preserving the interwoven meshwork patterns of ESM fibers. Thus, a bottom-up assembly of hierarchical SnO_2 which inherits the original configuration of ESM brings into effect the direction of the glycoprotein macromolecule chains (Fig.2.13). The glycoproteins of eggshell membrane are demonstrated to direct the nucleation, growth, and assembly of nanocrystalline SnO_2, on the basis of the biomaterial functioning as the physical substrate, the chemical revulsant, and the capping agent during the templating synthesis. Notably, the small nanocrystallite and interwoven tubular characteristics of the obtained hierarchical SnO_2 nanomaterials show practical potential in fabricating semiconducting sensors at the most sensitive, size-controlled mode.

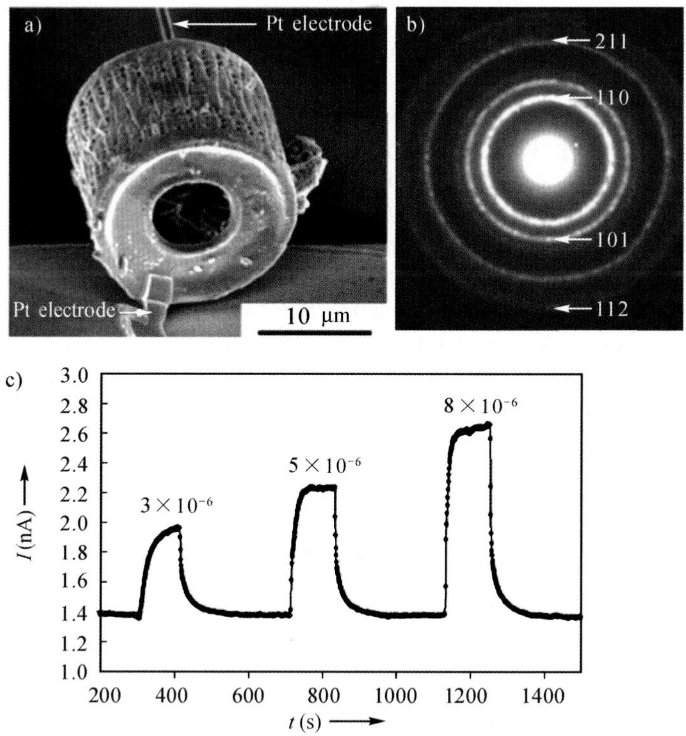

Fig.2.12. Deposition of conformal, continuous, and compact SnO_2 coatings on the surfaces of the valves of an *Aulacoseira* diatom through the combined use of dendritic surface-hydroxyl amplification and automated surface sol–gel processing. a) SEM image of the SnO_2-coated valves fabricated to be an NO sensor. b) SAED pattern of the cassiterite SnO_2 coating of the valve. c) Current (I) response on exposure to NO gas (3×10^{-6}, 5×10^{-6}, or 8×10^{-6} in argon). Reprinted from Weatherspoon et al., 2007. Copyright (2007), with permission from Wiley

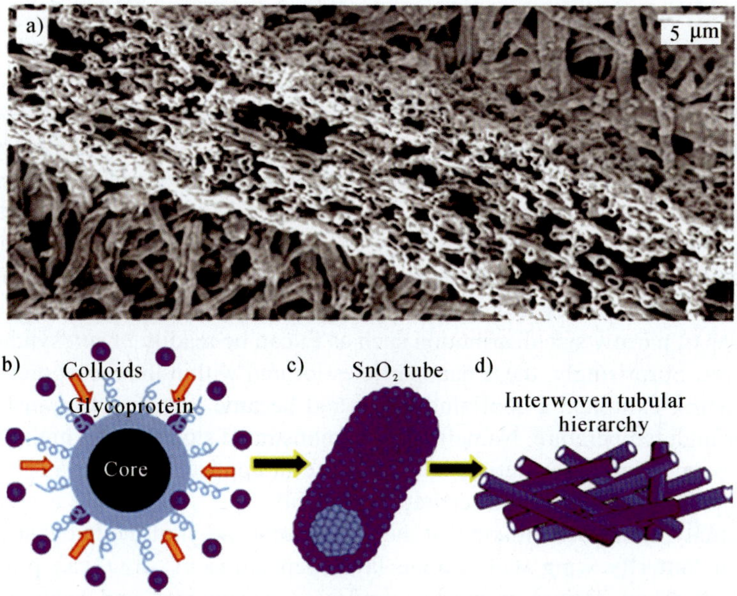

Fig.2.13. FE-SEM image of SnO_2 and schematic of the synthesis of hierarchical intertextures. a) Overview of the cross-section and the surface of the SnO_2 networks. Schematic illustration of the biogenic synthesis of SnO_2 hierarchical intertextures; b) cross-fractured shell membrane fiber displaying its glycoprotein biomolecules interacting with Sn colloids during the soaking reactions; c) SnO_2 tube assembled by nanocrystallites after the removal of the template ESM; d) interwoven SnO_2 meshwork yielded by nanotube arrays. Reprinted from Dong et al., 2007d. Copyright (2007), with permission from the American Chemical Society

2.2.4 Alumina Nanomaterials

Aluminum oxide is a well-known electrical insulator with a relatively high thermal conductivity. Also, its hardness makes it suitable to be used as an abrasive and scaffold to support functional materials. For example, TMV is coated with alumina thin films by the ALD method, giving highly stable nanotubes which inherit the original template morphology (Crowley et al., 2003).

Porous alumina is of much interest and importance due to its wide applications in adsorption and separation as well as catalysts and catalyst supports; and to obtain such material, biotemplates provide an effective pathway. Porous alumina ceramics with unidirectionally oriented pores by coating cotton fibers with alumina slurry was reported (Zhang et al., 2001). To fabricate more complex structured porous alumina materials, aluminum-vapor infiltration is employed to react with biocarbon templates (Rambo, Sieber, 2005). Rattan tissues were first pyrolyzed in an inert atmosphere. The resulting highly porous chars were then infiltrated with aluminum vapor and reacted to form Al_4C_3. Finally, oxidation and a sintering process converted the Al_4C_3 template into Al_2O_3. Through three-step high-temperature processing, highly porous biomorphic Al_2O_3 ceramics were obtained with hierarchical micromorphologies and well-oriented pore structures.

More easily, hierarchically-ordered porous γ-alumina was fabricated using *Pueraria lobata* as template by an immersion-fuming-calcination process (Li, He, 2006). *P. lobata* is a deciduous twining vine that spreads rapidly and covers everything in its path with a dense tangle of hairy stems and large trifoliate leaves. And its stem possesses a hierarchically-ordered porous architecture. The resultant replica precisely inherited all-level morphological features including those on both macro- and meso-scales of the template, like honeycomb parenchyma cells, and contained abundant mesopores. This exhibits the important advantage that small metal nanoparticles of narrow size distribution such as Pt can be readily *in situ* synthesized into the pores. Surprisingly, these nanoparticles located within the mesopores of the γ-alumna matrix exhibited a confining effect and became less easy to transfer and grow under high temperature. Namely, they demonstrated significantly high thermal stability. These γ-alumina materials have promise in applications such as catalysts, catalyst supports, absorbents, and separation materials.

Interestingly, it has been found that the precise replication of the initial structural hierarchy of butterfly wing with atomic layer deposition of Al_2O_3 can provide a direct way to retain optical properties similar to waveguides and beam-splitters (Parker, Townley, 2007). Butterfly wings are not only beautiful natural artworks but also highly functional template substrates. The varieties of fascinating colors exhibited are usually contributed by pigments and the presence of periodic submicrometer structures. The typical dimensions of butterfly wing scales that are about 150 μm long and about 50 μm wide demonstrate an extremely complicated morphology consisting of aligned lamellas that are in turn arranged into highly ordered net-like structures forming identical interspacing, which provide attractive template substrates for design and fabrication of advanced functional materials. A polycrystalline Al_2O_3 shell structure with precisely controlled thickness was obtained with optical properties such as the existence of a photonic band gap memorized. Compared to artificial photonic sensors, the resulting replicas showed higher selectivity, achieving a large difference in reflectance spectra in the vapors of solvents with similar polarities and refractive indices (including water, ethanol and methanol) without compromising their sensitivity (1×10^{-6}~2×10^{-6}), which remained in the same range as the artificial systems. This butterfly wing templated artificial material is not only a replica of the unique sophisticated structure and morphology, but also a combination of structure-related optical properties and enhanced sensing selectivity.

2.2.5 Zirconia Nanomaterials

Zirconium dioxide has attracted increasing interest in the field of heterogeneous catalysis due to its redox properties as well as acidic and basic character (Suh, Park, 2002). Its high chemical stability also favors its applications as a catalyst support, adsorbent, chemical sensor, and structural ceramic (Crepaldi et al., 2001; Vallet-Regí et al., 1997). The sol–gel coating procedure was applied for the zirconia coating of an ESM template, giving hierarchically ordered thin films with a macroporous network structure comprising crystalline zirconia tubes which are beneficial for its applications. While the resultant zirconia material exhibits the macroscopic morphology of a film about 15 μm thick, the film possesses a microstructure of macroporous networks composed of interwoven zirconia microtubes with diameters less than 1.0 μm. The tube walls consist of tetragonal zirconia nanocrystals with an average crystallite size of about 6 nm, which is formed during the calcination process carried out at 873 K to remove the natural template substrate. Once the calcination temperature was raised to 973 K, significant fusion between neighboring zirconia tubes accompanying a tetragonal-to-monoclinic phase transformation of zirconia was observed. The combined benefits of macropores, microtubes, and mesopores can potentially be obtained in one film, making this material attractive for numerous applications including catalysis, separation, and sensors.

2.2.6 Zinc Oxide Nanomaterials

The semiconductor zinc oxide (ZnO) has widespread applications, such as catalysts, piezoelectric transducers and actuators, microsensors, and photoelectrochemical cells, due to its outstanding chemical and physical properties, which mostly depend on the microstructures of the ZnO materials, including crystal size, orientation, and morphology. Furthermore, biotemplate synthesis provides a promising and feasible pathway for design and fabrication of ZnO materials with tailored structures and morphologies. It was reported that zinc oxide materials are endowed with a porous character and an interwoven meshwork pattern using an ESM template (Dong et al., 2007c).

Also, hierarchically periodic ZnO material was obtained by the butterfly wing scale templating procedure (Zhang et al., 2006b). By successive treatment with hydrochloride and sodium hydroxide, this relatively inert butterfly wing is activated, and bio-inspired tubular ZnO materials with structures of micrometer dimensions are obtained by the surface sol–gel process. The biomorphic three-dimensional porous structures maintained the microstructural features of the original butterfly wing scale morphology down to the micrometer level. Meanwhile, the resulting nanopores that exist on the walls are adjustable. Since ZnO has many functional applications, these replicas can serve as building blocks to construct micro- and nano-scale devices, such as UV light-emitting diodes, high-efficiency photonic devices, and photonic crystals.

Furthermore, *Spherobacterium* is employed as a template to prepare hollow zinc oxide spheres by an easy hydrothermal method (Fig.2.14) (Zhou et al., 2007). As a bio-candidate, bacteria exhibit distinct advantages such as being economical, environmentally-friendly, and safe, as well as the variety of well-defined morphologies controlled at the micro- or even nano-scopic level. Furthermore, there exist abundant functional groups on the cell walls which are able to bind metal cations or polar molecules through coordination or electrostatic interactions; no surface modification or activation steps are required, leading to a simplified process. Therefore, this bacteria-templating method can be extended to the preparation of various hollow structures such as hollow nanorods, nanotubes, nanohelixes, nanocables, diplo-spheres, chain spheres, and other kinds of three-dimensional nanostructures with diversity of morphologies that other biotemplates usually lack. To encapsulate the template bacteria with ZnO, triethanolamine molecules are first absorbed onto the cell walls combining with functional groups of carboxyls, hydroxyls, and phosphoryl of the cell walls, and further modify the bacterial surface. Then, ZnO clusters formed are inclined to grow by rapidly colliding with other ZnO clusters according to the ripening and aggregation theory (Spanhel, Anderson, 1991). As a result, ZnO covers the cell walls of bacteria. The resultant hollow spheres are likely to find applications in adsorption, gas sensors, optical devices, and photonic crystals. This economical, green, and convenient biotemplating strategy would open up possibilities for extensive study of the physical and chemical properties of these hollow structures and extend their application potentials in such areas as industrial catalysis, separation technology, environmental protection, electrochemistry, membranes, sensors, and optical devices.

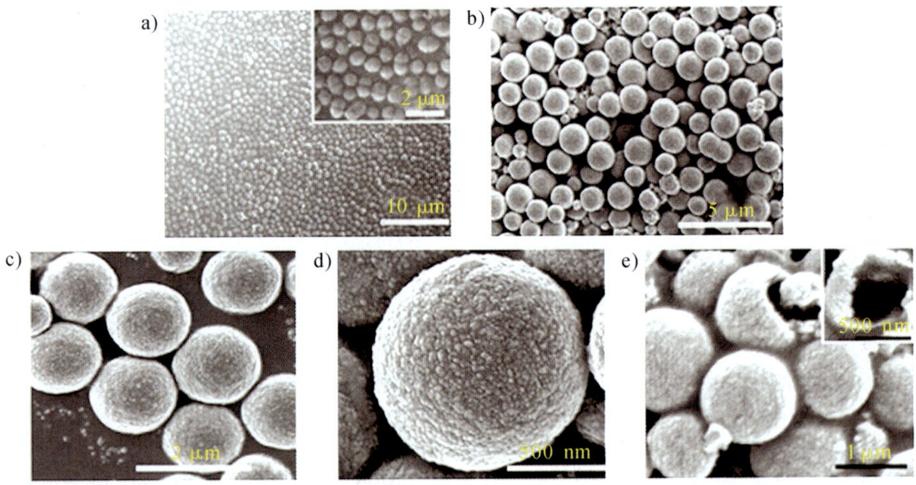

Fig.2.14. FE-SEM images: a) initial *Str. theromophilus* template, with the inset of higher magnification; b)~d) bacteria/ZnO core-shell spheres observed under various magnifications; e) ZnO hollow spheres after removal of bacterial templates by calcination at 873 K, with the inset of an individual broken hollow sphere. Reprinted from Zhou et al., 2007. Copyright (2007), with permission from Elsevier

2.2.7 Other Examples

In the typical sol–gel coating process, inorganic sol particles undergo surface-preferred gelation to form an inorganic coat on the templates. However, examples of biological substrates suitable for sol–gel coating are rare, probably because the commonly anionic surface at moderate conditions is not ideal. Interestingly, ashes obtained from the gills of mushrooms, silk fibers, and spider silk were found to be miniaturized replicas of the original materials, whereas ashes from dog hair and human hair were tubes. With interface-selective sol–gel polymerization (Kim, Jung, 2002), these materials were successfully coated with various inorganic materials regardless of the charges on them (Kim, 2003). Calcining coated template substrates yielded small structures composed of corresponding ash and coated inorganic materials such as silica, titania, copper oxide, aluminum oxide, and iron oxide. Fully calcined native specimens and as-coated samples appeared to shrink differently. Regardless of ash formation, biological materials are useful as templates for design and fabrication of small structured inorganic materials. The fact that ash replicas were smaller than the small original materials is also advantageous to obtain small complicated structures. The resulted inorganic small materials precisely memorized the original structures and morphologies, and can be used as building blocks in fabricating more complicated larger structures or as templates to prepare structures consisting of different materials.

2.3 Metallic Materials

Materials of single metals usually demonstrate unique properties. When such nanostructured materials meet complex structured natural substances, highly functional materials are created. The growth of metal on biomolecular templates is a promising route to producing complex metal nanostructures.

2.3.1 Nanostructured Gold

Mirkin and co-workers demonstrated for the first time that diatom silica walls could be chemically programmed to interact with inorganic nanoparticles (NPs) such as Au (Rosi et al., 2004). In those studies, a piranha etch solution of sulfuric acid/hydrogen peroxide mixture was used to digest the organic components of the diatoms and further activate the cell walls for subsequent aminosilane functionalization. The amino-functionalized diatoms were then reacted with ssDNA and acted as templates for the directed organization of Au nanoparticles functionalized with the cDNA. The nanoparticles formed a near-monolayer following the surface morphology and shape of the diatom template. Metal replicas identical to the respective diatom templates in nanostructure surface features as well as even sub-200 nm pore structures and other sub-100 nm topological features preserved were prepared (Payne et al., 2005). In addition, Au membranes with complex three-dimensional morphology that represent the exact negative of the initial porous diatom frustules which were used as masks for the evaporation of Au films were also fabricated (Losic et al., 2006).

Apart from these *in situ* metallization process, including palladium decorated ones (Richter et al., 2000; Richter et al., 2001; Ford et al., 2001), *ex situ* fabrication processes can also be used and exhibit different opportunities. In the reported *ex situ* approach, lysine-capped gold colloidal particles were decorated with DNA by electrostatic interaction between the positively charged particles and the negatively charged phosphate groups of the DNA (Kumar et al., 2001; Sastry et al., 2001). Actually, the negatively charged tris(hydroxymethyl)-phosphine-capped gold nanoparticles (THP-Au NPs) can also bind densely to calf thymus DNA, which enables DNA template synthesis (Harnack et al., 2002). DNA was first immobilized on a silicon substrate using O_2-plasma treatment to enhance immobilization, and spin-coating to elongate the molecules, followed by subsequent THP-Au NP sol treatment. Significantly, the mean distance between neighboring Au nanoparticles can be controlled by the length of the template DNA molecules. The efficiency of templating can be controlled by the solvent, since it affects the particle-particle and particle-DNA interactions. And the gold nanoparticles are enlarged with electroless gold plating solution in the end. The as-prepared nanowires show electrical conductivities *ca.* 1/1000 that of bulk gold. This metallization process is much faster than *in situ* approaches. Besides the individual gold-decorated nanowires, assemblies of gold nanoparticles with well-aligned and long-range order by using well-stretched DNA templates were reported (Nakao et al., 2003).

The skeletal plates of echinoids, sea urchins, were used to direct the formation of porous gold structures with nearly regular 15 μm channels (Payne et al., 2005). The skeletal plates and the spines of echinoids are mainly a low/high magnesium calcite and each behaves as a single crystal. The microstructure exhibits a unique fenestrated structure of interconnected trabeculae and well-controlled pore size. These highly uniform distributions of pores in very regular structures of the natural substance are faithfully mirrored by the porous gold product. The gold particles produce a coating on the calcite, rather than filling the entire pore structure, resulting in a double-sided surface. The materials presented here may find application as light-weight structural ceramics, for ordered macroporous materials with pore dimensions comparable to optical wavelengths can display unique optical properties. Furthermore, it can also be applied as catalyst supports, particularly in situations where the traditional materials with smaller pore sizes result in unacceptable pressure drops in the catalysis process.

On the other hand, the spherical *Chilo iridescent* virus (CIV) has also been used as biological templates for functional gold structure fabrication (Radloff et al., 2005). CIV consists of a layered structure that has a dsDNA-protein core surrounded by a lipid bilayer, which is in turn surrounded by an inner-outer capsid shell. The formed fibers are greater than 35 nm in length and protrude outward from the surface. By careful control of the ionic strength and pH of the solution, which affect the noncovalent interactions of the gold nanoparticles with the fibrils, gold nanoparticles are seeded around the viral capsid. In addition, gold nanocrystals (6 nm average diameter) are synthesized directly on the histidine-rich peptide templates by the *in situ* reduction of $ClAu^{3-}$ with $NaBH_4$, resulting in metallic nanowires (Djalali et al.,

2002). Similar to peptides, lipids can be directed to self-assemble into a variety of structures by fine-tuning the composition of the lipid molecules and by carefully controlled synthetic conditions. Certain types of structures, especially lipid tubes, have been used as biotemplates for the fabrication of metallic cylinders using metallization (Fig.2.15). More importantly, charged lipid tubules can also serve as effective templates for the fabrication of three-dimensional architectures and even novel helical structures. After these tubules were covered by alternating layers of anionic/cationic polymers, silica particles as small as 45 nm were adsorbed in a regular order (Lvov et al., 2000).

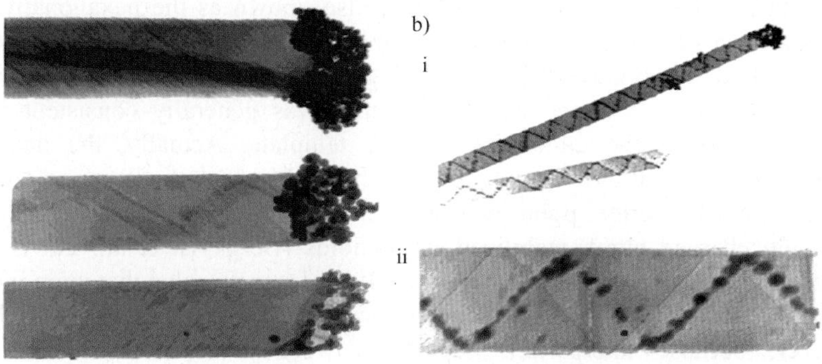

Fig.2.15. TEM images of lipid templated Au NPs at a) the caps of the tubes or b) in helical structures. Reprinted from Browning et al., 1998. Copyright (1998), with permission from the American Chemical Society

Protein fibrils can serve as scaffolds for the formation of metallic nanowire structures (Behrens et al., 2006a). Willner et al. proposed the design of gold nanowires based on the use of actin filaments as metallic nanotransporters on a myosin-modified surface, an example of a protein motor system (Patolsky et al., 2004). The design of actin-Au nanoblock patterned nanotransporters is envisioned to transport and release substances adsorbed to the gold elements at localized targets (Mertig et al., 1998; Pum et al., 2004; Sára et al., 2005; Sleytr et al., 2003; Sleytr et al., 1999).

S-layers, the outermost structural component of many bacteria, are two-dimensional crystalline arrangements of proteins or glycoproteins that feature a highly repetitive surface structure with nanometric unit cell dimensions, and display a variety of different lattice symmetries, including oblique, square, or hexagonal arrays with identical pore dimensions in the range of 2~8 nm diameter (Sára et al., 2005). The physical and chemical properties originating from such highly repetitive structures make S-layer lattices particularly suitable for the biotemplating of molecules and nanoparticles onto the surfaces with a series of chemical and physical approaches, including metal vapor deposition/argon ion milling (Panhorst et al., 2001), wet-chemical deposition followed by electron beam irradiation (Wahl et al., 2001), and site-specific assembly of presynthesized particles (Hall et al., 2001).

To enhance the immobilization of gold nanoparticles on the protein template, Dieluweit et al. introduced thiol groups into the primary structure of the S-layer of *B. sphaericus* (Dieluweit et al., 1998). Square arrays of Au particles with a 12.8 nm repeat distance were fabricated upon exposure to a tetrachloroauric (III) acid solution. Isolated gold nanoparticles of 4~5 nm were formed in the pore regions, resembling the morphology of the pore structure itself.

An alternative strategy for the fabrication of well-ordered arrays of nanoparticles is to use presynthesized colloids. The repetitive surface features of the S-layer from the positive bacterium *Deinococcus radiodurans* was used to direct the periodic assembly of negatively charged gold nanoparticles with diameter about 5 nm (Hall et al., 2001). The *D. radiodurans* S-layer, also known as the hexagonally packed intermediate (HPI) layer, comprises a hexameric protein core unit with a central pore surrounded by six relatively large openings that is called the vertex point. The interparticle center-to-center spacing (18 nm) was generally consistent with the lattice constant of the underlying S-layer template. Actually, the nanoparticle binding takes place at the vertex regions of the HPI layer, but adsorption tends to be at every second vertex point due to the interparticle repulsion forces of the negatively charged citrate-stabilized Au colloids (Bergkvist et al., 2004). Further studies of gold nanoparticle binding to the HPI S-layer revealed that upon increasing the ionic strength of the nanoparticle solution, ordered packing was still observed. However, because interparticle repulsions were less prominent under these conditions, adsorption of nanoparticles occurred at virtually every available vertex point, resulting in the formation of a honeycomb-like pattern of nanoparticles extending throughout the HPI monolayer sheet (Fig.2.16).

a)

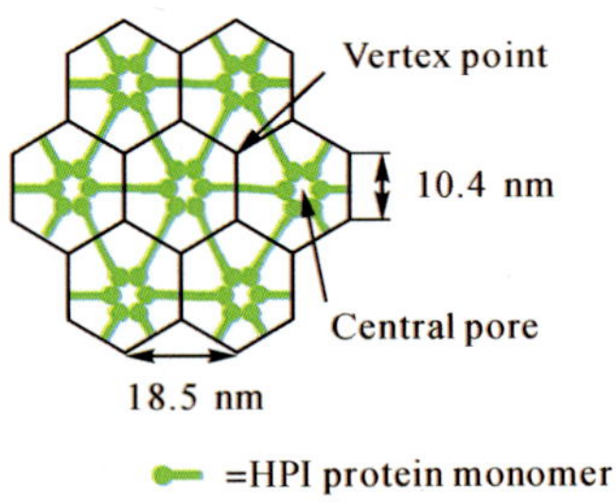

b)

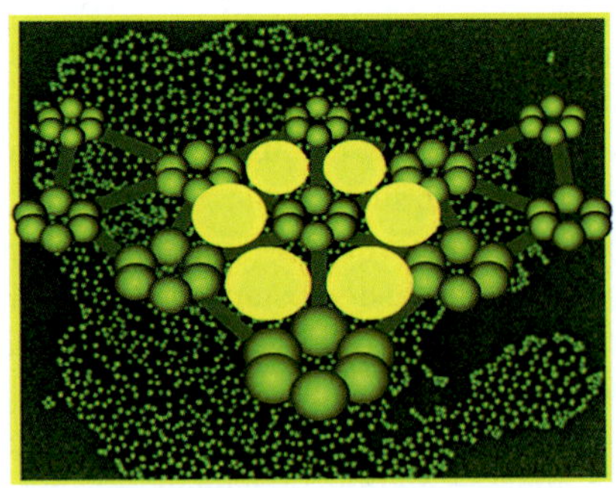

Fig.2.16. a) Schematic depiction of the structure and hexagonal symmetry of the HPI S-layers. b) SEM image (shown as background) of a honeycomb-like pattern of Au NPs adsorbed on the S-layer upon addition of 25 mM NaCl. Adsorption of NPs at the vertex point of the S-layer are shown in the cartoon representation. Reprinted from Sotiropoulou et al., 2008. Copyright (2008), with permission from the American Chemical Society

Recently, Keith et al. presented a protocol for the high yield preparation of stable colloidal suspensions of TMV/gold nanoparticle nanorods that exhibit uniform metallization, monodispersity in nanoparticle size, and small interparticle spacings (Fig.2.17). The procedure involves bio-mediated sequestration of tetrachloroauric acid ($HAuCl_4$), template-directed nucleation of gold clusters, and surface-mediated growth of metallic nanoparticles. The sizes and interparticle spacings of the densely packed gold nanoparticles are determined by the number of reaction cycles used. Novel steps, including addition of ethanol after the first addition-reduction cycle and wrapping of the TMV/Au nanoconstructions with poly-L-lysine, were introduced to ensure high homogeneity and stability in the nanowire suspension. These results not only demonstrate key improvements in the metallization of TMV but also provide a general protocol that could be applied to a wide range of bioderived templates such as DNA duplexes, lipid tubules, F-actin filaments, bacterial pili, and peptide fibers. Such materials have wide-ranging potential as novel bioinorganic components for integration into nanoelectronic devices and circuitry.

a)
b)

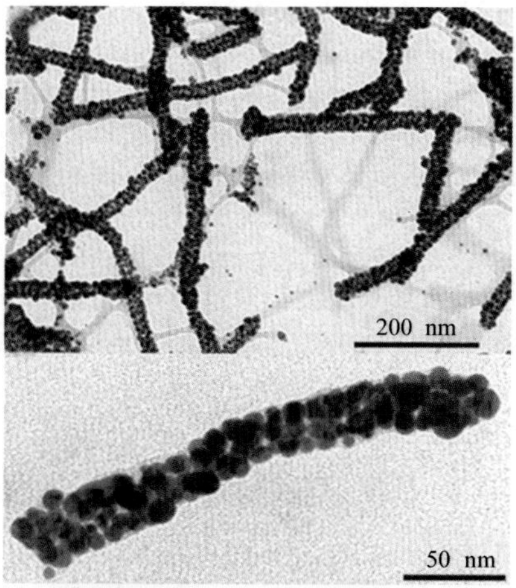

Fig.2.17. TEM images of as-synthesized Au NP-coated TMV nanowires. a) Low magnification image; b) isolated virion coated with a densely packed array of Au NPs. Reprinted from Bromley et al., 2008. Copyright (2008), with permission from the Royal Society of Chemistry

2.3.2 Nanostructured Silver

DNA can not only act as template substrate, but also guide the assembly of metal nanoparticles to form tailored nanostructures. Silver nanoparticles were first used to transcribe a single biopolymer strand by Braun et al. (Braun et al., 1998). In this case, a single strand of derivatized DNA was bridged between two gold electrodes, followed by reduction of the latter with hydroquinone to exchange the DNA Na^+ counterions for Ag^+ ions. In this way, small Ag particles were formed on the DNA template. Subsequently, more silver was deposited on this DNA-Ag strand by standard photographic procedures, resulting in a thin wire (length: 12 mm; diameter: about 100 nm) built up of small silver grains (30~50 nm) which presented interesting hysteric I-V curves. Thinner wires (25 nm) were also produced by this method, but the silver coating was shown to be discontinuous because some gaps existed between the silver grains. In later other work, one-dimensional metallic, semiconducting, and magnetic nanoparticle arrays as well as nanowires with diameters as small as 15 nm and displaying a variety of physicochemical properties such as metallic (Patolsky et al., 2002), fluorescent (Stsiapura et al., 2006), magnetic (Gu et al., 2006), and even binary semiconductors (Dong et al., 2007a) were successfully reported (Kinsella, Ivanisevic, 2007) with good control over the dimensions, crystallinity, and even chirality (Shemer et al., 2006).

Peptides, linear assemblies comprising 2~30 amino acid residues, can be designed to self-assemble into a wide variety of structures (Patolsky et al., 2004; Gazit, 2007; Behrens et al., 2004). With appropriate design, peptides can be self-assembled into tubular structures, ordered fibrillar structures, or even nanospheres. Moreover, a cyclic octapeptide with alternating L and D amino acids can self-assemble into nanotubular structures that further assemble into crystalline arrays. Along similar lines, linear hepta- and octa-peptides can self-assemble into tubular structures. Therefore, peptides have emerged as easily amenable building blocks for designing and creating nanotubular structures via self-assembly processes. Indeed, 20-nm Ag nanowires have been successfully cast using the inner hollow channel of such structures of peptides (Reches, Gazit, 2003). And further proteolytic degradation of the initial template allowed the isolation of discrete robust nanowires (Fig.2.18).

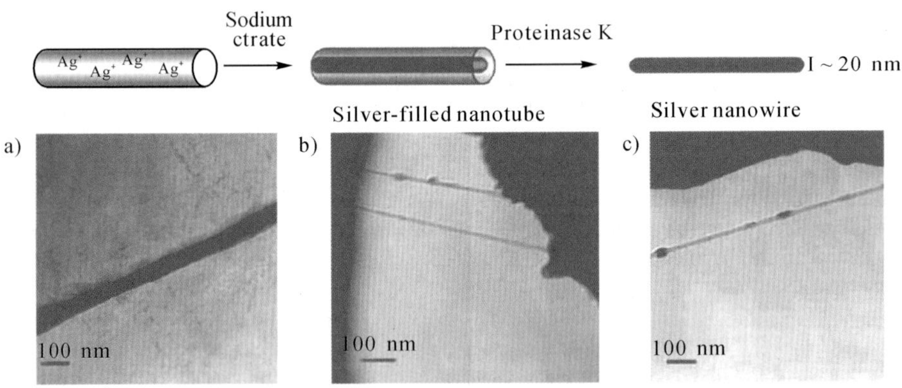

Fig.2.18. Casting of silver nanowires with peptide nanotube templates. a) The nanowires are formed by the reduction of silver ions within the tubes first, followed by enzymatic degradation of the peptide mold. b) TEM analysis of peptide tubes filled with silver nanowires. c), d) TEM images of silver nanowires that were yielded after proteolytic lysis of the peptide mold. Reprinted from Reches, Gazit, 2003. Copyright (2003), with permission from AAAS

Biomacromolecules other than DNA can direct coating materials in a different way. For instance, inorganic-organic hybrids were formed under the direction of silk fibroin fiber (SFF) biomacromolecules through a biotemplate redox approach at room temperature (Dong et al., 2005). The reductive amino acid tyrosine of SFF mainly provided both reduction and location functions under alkaline conditions (Selvakannan et al., 2004). This pH-dependent reducing ability of tyrosine arises due to ionization of the phenolic group of tyrosine at high pH, by which electrons transfer to silver ions, then to quinine, resulting in reduction of Ag(I) ions to Ag(0). Finally, highly stable silver polycrystalline grains of 5~10 nm diameter were generated on the SFF substrates. Thus, small-sized but well-crystallized silver nanocrystallites could associate with the biosubstrate SFF to be organized at specific locations of the biofibers with subtle hierarchy.

Actually, most natural substances are organic materials, and thus natural biological templates provide existing carbon sources instead of being removed. Carbon materials are thermally and mechanically stable, chemically inert, of low density, and sometimes highly porous, and thus are extensively utilized in various areas (Sakintuna, Yurum, 2005). Furthermore, they can act as substrates to support noble-metal nanoparticles, oxide nanoparticles, or semiconductor quantum dots (QDs), leading to functional carbon-based nanocomposites with specifically tailored properties. Regular, plump spherical carbon microspheres with supported silver nanoparticles were fabricated with pollen grains as the starting material as well as the carbon source (Wang et al., 2005a). Pollen grains are first rinsed in $AgNO_3$ aqueous solution due to the complexation and reduction functions of the active groups in the pollen for silver ions, followed by preoxidization in air at 573 K and subsequent carbonization in nitrogen at 873 K to carbonize the organic substrates. Here, preoxidization at the appropriate temperature is essential because the spherical structure of the pollen grains can be fixed and retained through cross-linking reactions of the components in the pollen with the aid of oxygen, or only shrunken carbon microparticles can be prepared.

2.3.3 Nanostructured Platinum

Michael et al. found that DNA bases have N bonding sites which can bind with Pt(II) complexes. The Pt(II)-activated DNA chains possess a high electron affinity that form stronger Pt-Pt bonds upon water substitution, igniting further Pt modification and forming metal cluster necklaces. The properties of DNA as a metallization template can be adjusted by controlling the activation time. Notably, if activation times are long enough, DNA can provide selectively heterogeneous nucleation sites for the formation of metallic platinum (Mertig et al., 2002).

A different approach involves the metallization of the native S-layer of *Sporosarcina ureae* for the production of Pt nanoparticles (Mertig et al., 2001). With this procedure, well-separated roughly spherical Pt(0) particles (~1.8 nm) with a uniform diameter distribution were formed and were spatially aligned along the tetragonal crystalline structure of the S-layer protein template.

2.3.4 Nanostructured Nickel

On the other hand, the mechanical properties of biotemplates can be improved by metallic coating. Nickel-coated bacteria fabricated by electroless chemical plating was measured by a powerful method for the study of mechanical properties at micro-/nano-scale — the nanoindentor, demonstrating that the related elastic modulus and hardness were significantly increased (Wang et al., 2007). Similarly, the gram-positive bacteria *Citeromyces matritensis* and *Bacillus* are used as template substrates to construct nickel-phosphorus shell nanostructures (Li et al., 2003).

In other studies, selective deposition of nanoparticles inside the hollow channel of virus resulted in the formation of nickel and cobalt nanowires several micrometers long and just a few nanometers (~3 nm) in diameter (Knez et al., 2003).

2.3.5 Nanostructured Copper

To achieve better electrical properties, wire-like structured nanomaterial was fabricated using copper to coat DNA template (Monson, Wooley, 2003). Copper is presently used to create connecting wires in integrated circuits, so DNA-directed fabrication of metallic copper nanostructures provides significant advances for nanoscale integrated circuit manufacturing. Copper(II) in aqueous $Cu(NO_3)_2$ was electrostatically associated with DNA aligned on a silicon surface, followed by ascorbic acid treatment to form a metallic copper sheath around the DNA. The resultant copper nanowires are about 3 nm high. These may be valuable as interconnects in nanoscale integrated circuitry, and can be readily generated from DNA molecules on surfaces. In addition, a more complete coating can be achieved by repeating the Cu(II) and ascorbic acid treatment. It is also reported that copper nanotubes were yielded after removal of the flagellin protein template (Kumara et al., 2007b).

2.3.6 Nanostructured Metallic Arrays

Trent et al. used *S. shibatae* to guide the chemical synthesis of hexagonally packed metallic arrays (McMillan et al., 2005). The intracellular heat-shock protein TF55β from *Sulfolobus shibatae* can spontaneously assemble into an octadecameric double-ring cage structure, named a chaperonin, with nine subunits per ring and a 10-nm diameter core. The topochemical properties of TF55β were genetically rendered to remove a loop that occludes the central pore of the assembled chaperonin and a polyhistidine (His_{10}) sequence was first fused to its amino terminus for synthesis. With these modifications, the solvent-accessible cores of assembled chaperonins displayed 180 additional His residues, creating a region with enhanced affinity for metal ions, which was spatially constrained by the interior dimensions of the chaperonin. When incubated with Pd^{2+}, the chaperonin cores performed as sites for selectively igniting the chemical reduction of magnetic transition metal ions from precursor salts. This procedure resulted bimetallic metal (Ni-Pd or Co-Pd) arrays with lattice dimensions defined by the engineered chaperonin.

Ferritin, a 450 kDa iron storage protein found in both prokaryotic and eukaryotic organisms, consists of a roughly cage-like spherical polypeptide shell that is self-assembled by 24 protein subunits with an exterior diameter of 12 nm and an interior diameter of 8 nm, where there are eight hydrophilic and six hydrophobic channels with pore sizes between 0.3 nm and 0.5 nm. Inside the ferritin shell, iron ions form crystallites together with phosphate and hydroxide ions that result a particle similar to the mineral ferrihydrite, and each ferritin complex is capable of accommodating up to 4,500 iron atoms (Harrison et al., 1991). Unfortunately, these protein components do not naturally form planar arrays *in vivo* when used as biotemplates. Thus, Okuda et al. induced the artificial crystallization of ferritin molecules *in vitro* at a water-air interface instead, resulting in hexagonally close-packed two-dimensional arrays of iron nanoparticles (diameter about 5.8~1.0 nm) and indium nanoparticles (diameter about 6.6~0.5 nm) (Okuda et al., 2005). The highly uniform arrays of more than 1 μm^2 in area were obtained by transferring the crystal films onto a silicon wafer. Notably, the formation of the particle arrays required the use of an aqueous subphase, where a number of critical factors that affected the spreading of the ferritin protein solutions at the surface (*e.g.,* subphase density, surface tension, and divalent ion concentration) had to be carefully controlled. Interestingly, the unique cage-like structure can be used for reconstruction, resulting in novel proteins constraining various inorganic nanoparticles such as Co_3O_4 (Allen et al., 2003; Resnick et al., 2006), Fe/Co all oxides (Klem et al., 2007), Mn_3O_4 (Mackle et al., 1993; Mann, 1991; Meldrum et al., 1995), CoPt (Warne et al., 2000), Pd (Ueno et al., 2004), Ag (Kramer et al., 2004), CdS (Wong, Mann, 1996), CdSe (Yamashita et al., 2004), ZnSe (Iwahori et al., 2005), $Cr(OH)_3$ and $Ni(OH)_3$ (Okuda et al., 2003), In_2O_3 (McMillan et al., 2005), FeS (Mann, Meldrum, 1991; Douglas et al., 1995), $CaCO_3$, $SrCO_3$, and $BaCO_3$ (Li et al., 2007) as well as branched polymers (Abedin et al., 2009), exhibiting potential magnetic, catalytic, and biomedical sensing applications.

2.3.7 Complex Metallic Materials

Rather than single inorganic particles, complex inorganic materials are used to fabricate enhanced optical devices like fly eyes, which are composed of integrated optical units called ommatidia (Martín-Palma et al., 2008). These units are spherically arranged following a hexagonal pattern along a curvilinear surface, and each ommatidium is composed of a facet lens, a crystalline cone, a set of photoreceptors and surrounding pigment cells (Stavenga, 2002). Such a formed corneal area reduces the reflection of incident light compared to a planar and optically smooth surface, and thus maximizes light transmission. The surface of an ommatidium is textured at the nanoscale, which provides a substrate with functional structure at the micro- and nano-scales for biotemplate synthesis. Chalcogenide glasses, nominal composition GeSbSe, were used for replication with the conformal-evaporated-film-by-rotation technique due to their infrared optical properties as well as good chemical and mechanical durability, which are specifically suitable for the development of optical components and sensors. The long-range spatial features of the initial template were replicated on the micron scale, together with the much finer features at the nanoscale. As a result, the original optical functionality was faithfully preserved except for slight differences in optical spectra, which are attributed to the different frequency dependences of the ocular material and the chalcogenide glass used for the replication. The replicated sub-wavelength structures can be used as antireflection structures leading to increased photon trapping.

Belcher et al. demonstrated the use of substrate-specific peptide-modified bacteriophage as a universal template to control the patterning of semiconducting, metallic, oxide, and magnetic materials (Mao et al., 2004). The substrate-specific peptides allow the formation of ZnS and CdS nanocrystals at site-specific locations within the viral capsid structure in a highly controlled manner. Indeed, the engineered peptides not only can direct the nucleation of nanocrystals, but also play a key role in the production of nanocrystals displaying preferred crystalline orientations. Annealing of the nanocrystals led to the formation of single-crystalline nanowires of the same crystallographic orientation as the precursor nanocrystals. Also, the fabrication of highly oriented quantum dot nanowires as well as the possibility to assemble hybrid nanomaterials was demonstrated (Mao et al., 2003). In addition, virus-templated gold nanoparticles were applied for the subsequent nucleation and growth of cobalt oxide nanowires into two-dimensional architectures over large scales (Huang et al., 2005), indicating a good dispersion of the Au-Co hybrid material. Several other materials, like iron oxides (22-nm diameter), CdS (5-nm diameter) and PbS (30-nm diameter), were also successfully deposited onto the TMV surface under corresponding guidance (Shenton et al., 1999). In addition, linear tubulin assemblies were used for the template-directed deposition of Pd (Behrens et al., 2002; 2006b), FePt (Beherens et al., 2006b) via electroless deposition techniques.

2.3.8 Other Examples

Rather than leaving intact DNA templates after the fabrication processes, Joseph et al. demonstrate a practical pathway to fragment the template substrate at a certain order (Kinsella, Ivanisevic, 2005). In this case, positively charged Fe_2O_3 nanoparticles were aligned along the negatively charged DNA surface through weak electrostatic interactions. The resultant stretched, surface-immobilized DNA retained its recognition properties. So, the assemblies were able to be digested by defined restriction enzymes. With enzyme treatment, the initially long nanowire was fragmented into short wires with certain lengths. This technique is useful to fabricate higher order nanostructured devices.

However, it should be noted that the simple hybridization of single-stranded oligonucleotides is generally not sufficient for the tailoring and synthesis of more complicated types of structures and architectures. Hence, synthetic DNA molecules featuring branched junction motifs with "sticky ends" flanking junctions were designed to enable the self-assembly of DNA sequences into two-dimensional and three-dimensional architectures, such as lattices and grids (Fig.2.19) (Feldkamp, Niemeyer, 2006; Yan et al., 2003). In turn, these structures can be further used for the templated assembly of 5-nm nanoparticles into ordered periodic arrays with excellent control over the lattice spacings, where interparticle distances range from 15 to 38 nm (Zhang et al., 2006a). This distinguished ability to control the spatial organization of NPs using robust DNA motifs led to the synthesis of binary arrays of nanoparticles with various sizes (Wang et al., 2006). Recently, Aldeye and Sleiman reported the design of "dynamic" single-stranded and cyclic DNA templates that allow for geometrical modularity of the nanoparticle assemblies (Aldaye, Sleiman, 2007). With this approach, exceptional control is possible not only over the spatial positioning of each nanoparticle, but also on their geometrical assembly. The presented ability to perform write/erase functions further exemplified the unique possibilities afforded by DNA in material templating applications.

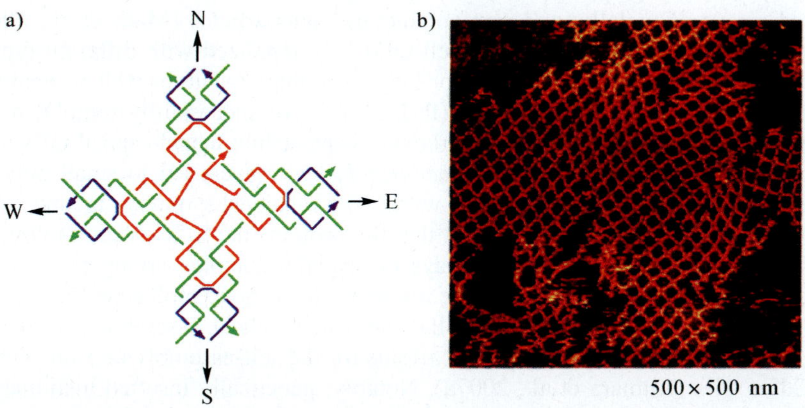

Fig.2.19. a) Schematic illustration of self-assembly of DNA nanogrids using a 4 × 4 DNA strand structure. b) Atomic force microscopic image of the resulting two-dimensional lattice (nanogrid). Reprinted from Yan et al., 2003. Copyright (2003), with permission from AAAS

2.4 Quantum Dots

Quantum dots (QDs), semiconductors whose excitons are confined in all three spatial dimensions, are particularly significant for optical applications due to their theoretically high quantum yield. They have been proven to perform like a single-electron transistor, showing the Coulomb blockade effect, and are suggested as implementations of qubits for quantum information processing. Protein fibers are a large class of structural biomaterials that play an essential role in the motility, elasticity, scaffolding, stabilization, and protection of cells, tissues, and organisms in biological systems (Scheibel, 2005). Due to their unique morphological properties and molecular recognition capabilities, fibrous proteins have been used as scaffolds for the templated assembly of semiconducting QDs and for the *in situ* deposition of various metallic species in a variety of interesting nanoarchitectures (Kumara et al., 2007a).

In 1997, Shenton et al. demonstrated that native *Bacillus stearothermophilus* and *Bacillus sphaericus* can be used to induce the *in situ* mineralization of CdS superlattices through a combination of wet and vapor phase chemical deposition processes (Shenton et al., 1997). In this case, self-assembled S-layers were exposed to a $CdCl_2$ metal-salt solution for several hours, followed by slow vapor-phase reaction with a reducing agent (H_2S) over a period of one to two days. The obtained CdS nanocrystals (*ca.* 4~5 nm) were mainly localized to the pore regions between the subunits of the S-layers and arranged in a periodic pattern that corresponded to the oblique and square lattice symmetries of the respective S-layers.

Mark et al. reported that it is possible to tailor, within certain limits, the particle size-surface interactions to complement the topochemistry of the protein lattice substrate by choosing appropriate NP surface ligands and further form self-organized, ordered arrays of metallic and semiconducting nanoparticles (Mark et al., 2006a). Various species of CdSe/ZnS core-shell QDs functionalized with different types of negatively or positively charged/short- or long-chain thiol ligands as well as dendrimer-encapsulated platinum nanoparticles (Pt-DENs) were successfully templated. In a striking case, the biotemplating of 7-carboxy-1-heptanethiol ligand-capped QDs on the thermacidophilic archaeon *Sulfolobus acidocaldarius* S-layers led to small polygonal clusters comprising three QDs arranged with 3-fold rotational symmetry around isolated QDs (Fig.2.20). These results suggest that the protein-like S-layer can biologically program the formation of uniform arrays of spatially complex arrangements of NP clusters without any additional specific interparticle bridging molecules.

A report shows the use of flagella, elongated helical assemblies of flagellin proteins that are up to 10~15 μm, as scaffolds for the self-assembly of 3 nm ZnS/Mn and CdTe QDs (Kumara et al., 2007a). Notably, genetically inserted histidine loop peptides in the flagella controlled the subsequent deposition of QDs in a regular fashion. Apart from leveraging the structural specificity, proteins with dynamic motility functions can add another avenue as biotemplates with applications in nanoelectromechanical systems as microconveyor belts for the sorting and delivery of nanomaterials. Muthukrishnan et al. functionalized CdSe/ZnS QDs with kinesin motors via a biotin-neutravidin linkage (Muthukrishnan et al., 2006). Here, the cellular nanoscale transportation machines, called motor proteins surfaces, were modified with kinesin, a type of cargo transporting protein motor, which guided the transport of kinesin-neutravidin-QD complexes, according to the authors. This approach has advantages for transporting larger, kinesin functionalized materials, such as beads92 and QDs. Differently, CdSe QDs were coupled to microtubule proteins using a biotin-streptavidin linkage (Bachand et al., 2004). In this case, QD assembly was confined to the central region within the microtubule proteins to allow the microtubule filament ends to interact with immobilized kinesins. Thus, the kinesin motor proteins were able to successfully transport the QD-MT composites

(Fig.2.21). To realize the full potential of molecular motor conveyor belts in nanoscale devices, fine control over transport directionality is the key point. To this end, a combination of chemical and topographical patterns is reported to selectively bind kinesin motors and subsequently guide the motility of MTs on the kinesin-modified surfaces (Hess et al., 2001; van den Heuvel et al., 2005).

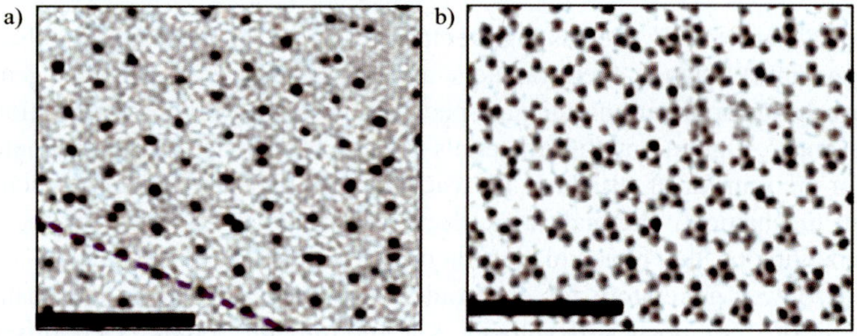

Fig.2.20. Bright-field TEM images of different nanoparticle patterns on unstained SAS S-layers after incubation in solutions of QDs functionalized with 7-carboxy-1-heptanethiol ligands. a) and b) show the particle adsorbed to the two different "faces" of the SAS S-layer. Bars = 100 nm. Reprinted from Mark et al., 2006a. Copyright (2006), with permission from the American Chemical Society

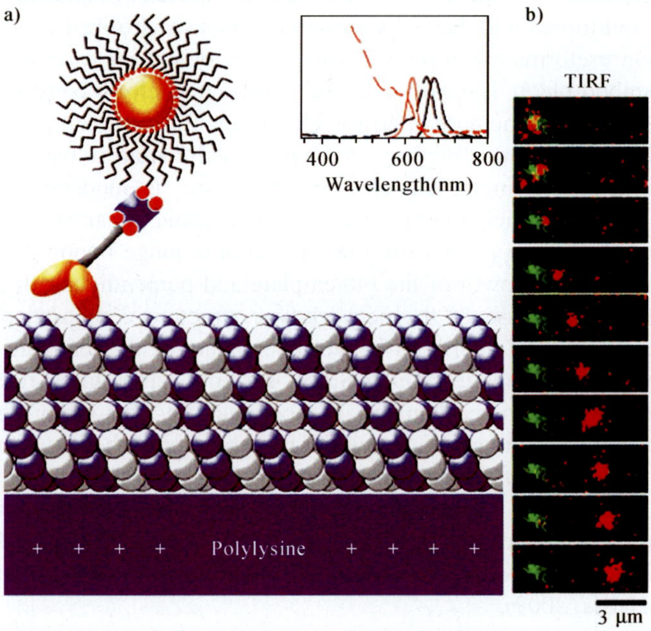

Fig.2.21. a) Schematic representation of a kinesin-neutravidin-QD complex moving along a surface-immobilized microtubule. b) Total internal reflection fluorescence (TIRF) image showing a QD-kinesin complex moving along an immobilized, fluorescently-labeled MT (elapsed time ~20 s). Reprinted from Muthukrishnan et al., 2006. Copyright (2006), with permission from Wiley

2.5 Silica Carbide Materials

SiC materials are widely used as high temperature electric heaters. They also exhibit advantages in resistance thermometers and in thermoelectric power generation, targeting various applications such as space technology, local communication, small batteries and refrigeration. SiC materials developed through the biomimetic route can retain the imprint of structural intricacies of the native templates, which exhibit excellent mechanical, thermal, and electrical properties. Because of the feasible maneuverability of the biotemplate structures, sophisticated designs of electrical heaters or power generators for electronic circuitry can be developed using such cellular Si/SiC composites developed via biomimetic routes. Debopriyo et al. reported the growth of Si/SiC ceramic composites with natural bamboo as the biotemplate as well as the carbon source (Mallick et al., 2007). Bamboo has several distinguishing structural features (Li et al., 1995). While it is a cylinder on the macroscale, it shows a nonlinear gradient structure comprised of vascular bundles oriented along the growth axis of the thin-walled cells on the mesoscale. And on the microscale, bamboo bust fibers are hollow tubes composed of concentric layers, and each layer is reinforced with helically wound cellulose microfibrils. After pyrolysis, skeletal carbon preforms retain the structural integrity and microcellular features of the native bamboo plant. The carbon of the pyrolyzed preform converts to SiC and the residual Si fills into the pore interiors. The resultant Si/SiC composite forms part of the complicated network structure, retaining the cellular array feature of the initial native template. Interestingly, the electrical resistivity is moderately anisotropic, whereas the thermoelectric power turns out to be isotropic. Meanwhile, the electrical conduction mechanism appears to follow the variable range hopping scenario both along the direction of growth of the biotemplate and perpendicular to it. It is quite apparent then that because of the network-type structure composed of strands of different dimensions and interfaces between remnants of the template and Si/SiC, structural disorder develops in cellular Si/SiC systems, which gives rise to a switch in the charge conduction mechanism. The conductivity anisotropy appears to be driven primarily by the unique microcellular morphology of the biotemplates, which can be exploited in many electrical applications. Other investigations on the electrical properties of biomorphic SiC ceramics at low as well as at high temperatures, are also reported. Natural templates in synthesizing Si/SiC systems are not limited to bamboo; biotemplates such as ESM, echinoid plates, and butterfly wings are also used (Maddocks, Harris, 2009).

Notably, on the basis of the pyrolysis mentioned above, bio-inspired porous carbon-rich materials can be obtained with further treatment. Ordered macroporous-mesoporous carbonaceous materials were prepared as a direct replica of the *Thalassiosira pseudonana* diatom by infiltration of the skeleton with furfuryl alcohol as carbon precursor (Pérez-Cabero et al., 2008). Through pyrolysis and subsequent removal of the silica component by chemical etching, the final carbon-rich material was obtained with the initial macropores of the diatom template and new hierarchical macro-mesopores, giving an organized array of carbon macrotubes (Fig.2.22). The use of a specific diatom as bio-template opens the possibility to synthesize accurate direct/inverse carbon replicas with well-defined and tailored morphologies and structures. The large variety of available diatoms with macropores that range from the nano- to the micro-metric domains can be considered as a wide catalogue of biotemplates to produce ordered macroporous carbon-based materials. The extension of classical application fields of the diatoms to hydrophobic media constitutes the evident utility of these materials.

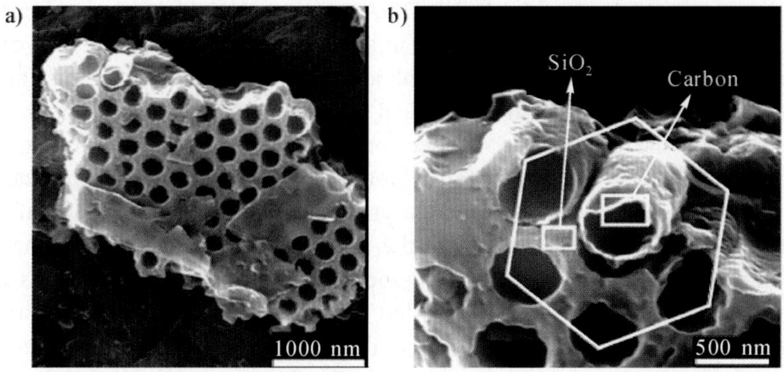

Fig.2.22. SEM images of C/Al-Tp after etching. Reprinted from Pérez-Cabero et al., 2008. Copyright (2008), with permission from Elsevier

2.6 Materials Fabricated by Organic Coating

The construction and evaluation of materials and devices with nanometer dimensions has attracted increasing research interest. Carbon nanotubes have emerged as important materials for nanofabrication, in both electronic devices (Collins et al., 1997; Tans et al., 1997; Shim et al., 2001) and sensors (Kong et al., 2000; Varghese et al., 2001; Chopra et al., 2002; Qi et al., 2003; Valentini et al., 2003). A key issue for the use of carbon nanotubes in nanofabrication is whether their placement is controlled at well-defined positions on surfaces. An intriguing approach for assembly of nanoscale materials is the use of a biological template to direct the positioning of components on a surface. With DNA template direction, selective localization of single-walled carbon nanotubes (SWNTs) is achieved (Xin, Wooley, 2003). Template γ-DNA was decorated with 1-pyrenemethylamine hydrochloride (PMA) and aligned on a silicon wafer surface, leading SWNTs to deposit onto the designed location. This approach offers significant potential to facilitate the construction of ordered arrays of nanometer dimension materials. Furthermore, the bottom-up construction of nanoscale electronic circuits can be realized under the combination of DNA-templated nanotube localization with the ability to form specific base pairs between oligonucleotide-coupled nanostructures and surface DNA.

Polyaniline is commonly synthesized by oxidizing aniline monomer not only chemically but also electrochemically in a strongly acid environment, required to initiate the polymerization reaction (Verghese et al., 1996; Liu, Freund, 1997). Although the harsh condition precludes its application to delicate biologically-based materials as templates, the environmentally benign horseradish peroxidase (HRP) enzymatic approach was reported to provide opportunities for the use of biological polyelectrolytes, such as DNA, as templates. DNA-templated water-soluble DNA/polyaniline complexes were produced (Nagarajan et al., 2001), in which the polyaniline was wrapped around the DNA matrix. Notably, it could reversibly control the secondary structure of DNA in solution, and furthermore, it avoided the possible highly branching reaction of polyaniline, which severely limits the degree of conjugation and the electrical and optical properties of the resulting polymers.

To improve the application in electronics, polyaniline nanowire arrays were fabricated on Si chips by using fully-stretched DNA as growth templates while applying the gentle HRP enzymatic polymerization approach (Ma et al., 2004). Since the pH significantly affects both the catalytic activity of the enzyme and the formation of alignment of the aniline monomer along the DNA templates, the pH of the polymerization was strictly controlled to around 4.0 to form continuous and conductive polyaniline nanowires. The conductivity of the nanowires is sensitive to acid-base doping and undoping processes, indicating potential sensing applications such as gaseous hydrochloride and/or ammonia sensors.

Moreover, virus-virus interactions such that a two-dimensional liquid crystalline layer was created on top of polyelectrolyte films under the principles of self-assembly were organized. The resultant material exhibits a potential for achieving high cycling rates with capacity remaining practically stable for up to 10 charge/discharge cycles and can be operated at equivalent or higher capacitance values, which are suitable for lithium-battery applications. The total power output of their systems can potentially be increased by simply assembling a number of alternating stacks of the material (Nam et al., 2006).

Although it is generally believed that soft biological organs are difficult to use as engineering technical templates, some biological substances can endure hard processing conditions. For example, cicada wings can endure high temperatures up to 473 K (Xie et al., 2008). They consist of chitin, a high molecular weight, crystalline polymer whose Young's modulus is as high as 7~9 GPa (Vincenta, Wegst, 2004), which is high enough to maintain the original wing surface structure during the replication process. Furthermore, a layer of wax present on the epidermis of cicada wings is a low surface tension material (Vincenta, Wegst, 2004), and can serve naturally as an anti-sticking coating during the replication process to avoid adherence. The transparent wing surface contains a quasi-two-dimensional, hexagonally ordered assembly of nano-nipples. These subwavelength nipple arrays endow the wing with the antireflection function associated with camouflage. They effectively introduce a gradual refractive index profile at the interface between wing and air, and reduce the reflectivity over broad angles or frequency ranges by a factor of about ten (Parker, Townley, 2007). With this "hard" functional biological organ, a polymethyl methacrylate replica was prepared, preserving not only the photonic structure of the original cicada wing surface, but also the anti-reflective property. In the typical process, a negative gold mold is first prepared directly from the bio-template of the cicada wing by thermal deposition, and then the gold mold is used to get the replica of the nanostructures on the original cicada wing by casting polymer. The nano-nipple arrays on the surface of cicada wings were precisely replicated to the surface of a PMMA film (Fig.2.23). This subwavelength anti-reflective PMMA film can be used as an "absorber" component, which may have applications in solar cells, high-power laser windows, and even prescription glasses.

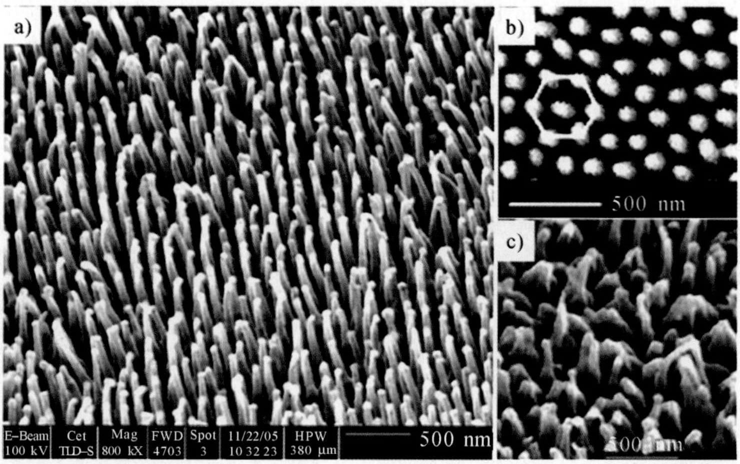

Fig.2.23. SEM images of the copied PMMA films with nano-nipple arrays on the surface from the negative Au mold. a) Large-scale perspective view. b) Higher-magnification top view showing a hexagonal pattern. While a) and b) were obtained after the PMMA film was heated at 363 K for 30 min, c) presents a perspective view after the film was heated at 333 K for 30 min. Reprinted from Xie et al., 2008. Copyright (2008), with permission from the IOP and IOP Publishing Limited

2.7 Other Natural Substance-derived Materials

The combination of bottom-up self-assembly and top-down lithographic approaches reveals another possible means to better control the placement of materials as well as provide a wider range of potential nanostructures and morphologies. With this strategy, a periodic array of nanometric holes (~10 nm in diameter) was created by Douglas et al. (1992) (Fig.2.24).

Typically, S-layer patches isolated from *S. acidocaldarius* (SAS) were first attached to a freshly-cleaved highly-oriented pyrolytic graphite substrate and shadowed at an angle with titanium to produce a mask for the subsequent ion milling. Similarly, a series of ordered metallic nanostructures were created with S-layers templates for thin-film lithographic processing (Winningham et al., 1998).

In a further extension of the protein masking/etching technique to more technologically relevant substrates, pattern transfer from an S-layer biomolecular nanomask to crystalline silicon (Si) substrates was achieved by an inductively-coupled plasma (ICP) etch process (Winningham et al., 2001). In this methodology, a so-called intermediate transfer layer (ITL) comprising a layer of a resist-like material such as nitrocellulose and polyimide, is first applied to the silicon substrate before deposition of the S-layer. The bionanomask pattern is first transferred to the ITL and then to the substrate. With proper ICP etch and/or metal lift-off control, template S-layers are deployed on an ITL of ultrathin (<10 nm) nitrocellulose to pattern Si substrates with either a two-dimensional ordered array of about 10 nm diameter holes, or, alternatively, a two-dimensional array of about 10 nm diameter metal dots (Ti, Pd, or Au). Both types of arrays show hexagonal symmetry and a lattice constant of 22 nm, consistent with the morphological structure of the protein bionanomask, with the exception of occasional defects.

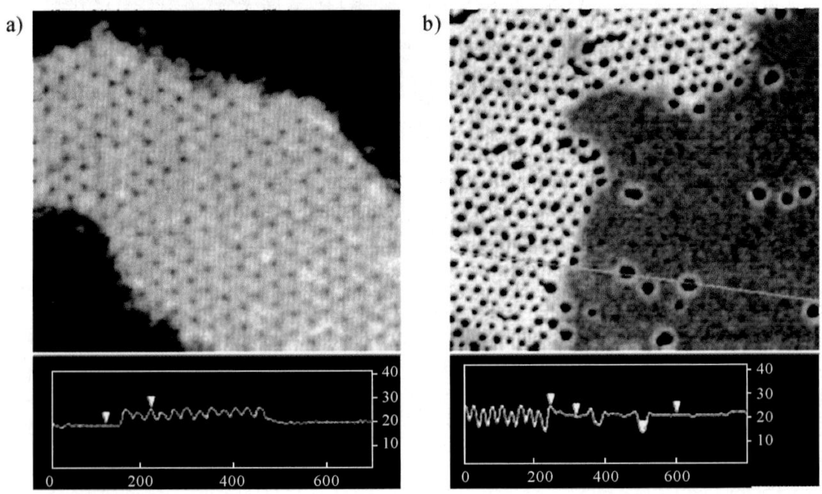

Fig.2.24. Transfer of S-layer-derived nanometer-scale patterns to highly oriented pyrolytic graphite surfaces by ion milling. a) AFM image of TiO_x-coated on-S-layer and off-S-layer areas before fast atom beam milling. The S-layer lattice constant of 22 nm serves to indicate the length scale of the image. The cross-sectional profile along the line in the AFM image is also shown in the lower column. b) AFM image of TiO_x-coated on-S-layer and off-S-layer areas after FAB milling. Reprinted from Douglas et al., 1992. Copyright (1992), with permission from AAAS

Using an ICP SiC_4 etch process accompanied by an etch mask generated from 5 nm Au nanoparticles adsorbed onto the HPI S-layer, 100 nm high silicon nanopillars were prepared (Mark et al., 2007). The resulting silicon nanopillar structures possess a unique structure that is 8~13 nm wide at the tip, 15~20 nm wide at half-height, 20~30 nm wide at the base, and 60~90 nm high, and they appear to lack any significant degree of translational ordering, suggesting that further studies are necessary to elucidate the optimal plasma processing parameters that lead to the generation of long-range ordered arrays of silicon-based nanostructures using S-layer protein templates.

2.8 Summary

In summary, biotemplate synthesis possesses notable advantages, including the sheer structural diversity of available biological species and materials, the unique complicated architectures and degree of complexity achievable, as well as significant structure-related properties. Together, these elements provide for the design and fabrication of a diverse range of novel materials with an unprecedented repertoire of structures and morphologies that extend beyond those of the currently available conventional artificial materials. Also, this strategy is potentially more cost- and time-effective compared with current techniques for materials research. In addition, the repetitive topochemical features and variety of functional groups that exist in many biological materials can be employed for the *in situ* synthesis and directed self-assembly of both organic and inorganic nanostructures under mild conditions without additional chemical treatment.

On the other hand, through rational genetic engineering and/or targeted chemical modification, biotemplating enables spatially precise modification at the molecular level. Taken together, biotemplate synthesis provides great diversity and an easy approach to the tailoring and preparation of a variety of functional materials with promising potential for various practical applications. Bio-inspired science and technology is at the intersection of biology, chemistry, physics, and materials science; and is a bridge to connect new-age nanotechnology and classic biological science. The full application and exploration of biological templates has only just ignited.

Actually, researchers are just beginning to grasp the effects of nanoscale morphologies and structures on the optical, chemical, and electrical properties of materials. Precise replication of natural three-dimensional biological structures at the nanometer scale, and further at the single molecule level with a certain guest material is still a fairly unexplored field. Also, the currently reported strategies still lack high yields and precise uniformity. In particular, large-scale fabrication is an issue in some cases because there are insufficient quantities of purified biological material, or because of a lack of long-range order in the final product due to intrinsic lattice/morphological defects in the biotemplate itself. Moreover, because the exact mechanisms by which biological entities form defined patterns and guide the growth of crystalline materials are not yet fully understood, biotemplate synthesis is usually conducted in a highly empirical manner. Clearly, there is a great potential for using biological materials to develop entirely novel systems that display superior characteristics, but further research is necessary. Ultimately, it is expected that the biotemplating approach will provide innovative pathways to create novel materials that possess complex micro- and nano-structures with greater speed, precision, and flexibility as well as at a lower cost than traditional manufacturing processes.

References

Abedin MJ, Liepold L, Suci P, Young M, Douglas T (2009) Synthesis of a cross-linked branched polymer network in the interior of a protein cage. J Am Chem Soc 131:4346-4354

Aldaye FA, Sleiman HF (2007) Dynamic DNA templates for discrete gold nanoparticle assemblies: control of geometry, modularity, write/erase and structural switching. J Am Chem Soc 129:4130-4131

Allen M, Willits D, Young M, Douglas T (2003) Constrained synthesis of cobalt oxide nanomaterials in the 12-subunit protein cage from Listeria innocua. Inorg Chem 42:6300-6305

Anderson MW, Holmes SM, Hanif N, Cundy CS (2000) Hierarchical pore structures through diatom zeolitization. Angew Chem Int Ed 39:2707-2710

Bachand GD, Rivera SB, Boal AK, Gaudioso J, Liu J, Bunker BC (2004) Assembly and transport of nanocrystal CdSe quantum dot nanocomposites using microtubules and kinesin motor proteins. Nano Lett 4:817-821

Bae A-H, Numata M, Hasegawa T, Li C, Sakurai K, Shinkai S (2005) 1D arrangement of Au nanoparticles by the helical structure of schizophyllan: a unique encounter of a natural product with inorganic compounds. Angew Chem Int Ed 44:2030-2033

Behrens S, Habicht W, Wagner K, Unger E (2006a) Assembly of nanoparticle ring structures based on protein templates. Adv Mater 18:284-289

Behrens S, Habicht W, Wu J, Unger E (2006b) Tubulin assemblies as biomolecular templates for nanostructure synthesis: from nanoparticle arrays to nanowires. Surf Interface Anal 38:1014-1018

Behrens S, Rahn K, Habicht W, Bohm KJ, Rosner H, Dinjus E, Unger E (2002) Nanoscale particle arrays induced by highly ordered protein assemblies. Adv Mater 14:1621-1625

Behrens S, Wu J, Habicht W, Unger E (2004) Silver nanoparticle and nanowire formation by microtubule templates. Chem Mater 16:3085-3090

Bergkvist M, Mark SS, Yang X, Angert ER, Batt C (2004) Bionanofabrication of ordered nanoparticle arrays: effect of particle properties and adsorption conditions. J Phys Chem B 108:8241-8248

Braun E, Eichen Y, Sivan U, Ben-Yoseph G (1998) DNA-templated assembly and electrode attachment of a conducting silver wire. Nature 391:775-778

Bromley KM, Patil AJ, Perriman AW, Stubbs G, S Mann (2008) Preparation of high quality nanowires by tobacco mosaic virus templating of gold nanoparticles. J Mater Chem 18:4796-4801

Browning SL, Lodge J, Price RR, Schelleng J, Schoen PE, Zabetakis D (1998) Fabrication and radio frequency characterization of high dielectric loss tubule-based composites near percolation. J Appl Phys 84:6109-6113

Chia S, Urano J, Tamanoi F, Dunn B, Zink JI (2000) Patterned hexagonal arrays of living cells in sol–gel silica films. J Am Chem Soc 122:6488-6489

Chopra S, Pham A, Gaillard J, Parker A, Rao AM (2002) Carbon-nanotube-based resonant-circuit sensor for ammonia. Appl Phys Lett 80:4632-4634

Collins PG, Zettl A, Bando H, Thess A, Smalley RE (1997) Nanotube nanodevice. Science 278:100-102

Cooper AI (2003) Porous materials and supercritical fluids. Adv Mater 15:1049-1059

Crepaldi EL, Soler-Illia G, Grosso D, Albouy P-A, Sanchez C (2001) Design and post-functionalisation of ordered mesoporous zirconia thin films. Chem Commun: 1582-1583

Crowley TA, Ziegler KJ, Lyons DM, Erts D, Olin H, Morris MA, Holmes J (2003) Synthesis of metal and metal oxide nanowire and nanotube arrays within a mesoporous silica template. Chem Mater 15:3518-3522

Culver JN, Dawson WO, Plonk K, Stubbs G (1995) Site-directed mutagenesis confirms the involvement of carboxylate groups in the disassembly of tobacco mosaic virus. Virology 206:724-730

Dennis JE, Carrino DA, Yamashita K, Caplan AI (2000) Monoclonal antibodies to mineralized matrix molecules of the avian eggshell. Matrix Biol 19:683-692

Dieluweit S, Pum D, Sleytr UB (1998) Formation of a gold superlattice on an S-layer with square lattice symmetry. Supramol Sci 5:15-19

Djalali R, Chen Y, Matsui H (2002) Au nanowire fabrication from sequenced histidine-rich peptide. J Am Chem Soc 124:13660-13661

Dong A, Wang Y, Tang Y, Ren N, Zhang Y, Yue Y, Gao Z (2002) Zeolitic tissue through wood cell templating. Adv Mater 14:926-929

Dong L, Hollis T, Connolly BA, Wright NG, Horrocks BR, Houlton A (2007a) DNA-templated semiconductor nanoparticle chains and wires. Adv Mater 19: 1748-1751

Dong Q, Su H, Cao W, Zhang D, Guo Q, Lai Y (2007b) Synthesis and characterizations of hierarchical biomorphic titania oxide by a bio-inspired bottom-up assembly solution technique. J Solid State Chem 180:949-955

Dong Q, Su H, Xu J, Zhang D, Wang R (2007c) Synthesis of biomorphic ZnO interwoven microfibers using eggshell membrane as the biotemplate. Mater Lett 61:2714-2717

Dong Q, Su H, Zhang D (2005) *In situ* depositing silver nanoclusters on silk fibroin fibers supports by a novel biotemplate redox technique at room temperature. J Phys Chem B 109:17429-17434

Dong Q, Su H, Zhang D, Cao W, Wang N (2007d) Biogenic synthesis of tubular SnO_2 with hierarchical intertextures by an aqueous technique involving glycoprotein. Langmuir 23:8108-8113

Douglas K, Devaud G, Clark NA (1992) Transfer of biologically derived nanometer-scale patterns to smooth substrates. Science 257:642-644

Douglas T, Dickson DPE, Betteridge S, Charnock J, Garner CD, Mann S (1995) Synthesis and structure of an iron(III) sulfide-ferritin bioinorganic nanocomposite. Science 269:54-57

Feldkamp Niemeyer CM (2006) Rational design of DNA nanoarchitectures. Angew Chem Int Ed 45:1856-1876

Fernandez MS, Araya M, Arias JL (1997) Eggshells are shaped by a precise spatio-temporal arrangement of sequentially deposited macromolecules. Matrix Biol 16:13-20

Ford WE, Harnack O, Yasuda A, Wessels JM (2001) Platinated DNA as precursors to templated chains of metal nanoparticles. Adv Mater 13:1793-1797

Fowler CE, Shenton W, Stubbs G, Mann S (2001) Tobacco mosaic virus liquid crystals as templates for the interior design of silica mesophases and nanoparticles. Adv Mater 13:1266-1269

Gallis KW, Landry CC (2001) Rapid calcination of nanostructured silicate composites by microwave irradiation. Adv Mater 13:23-26

Gazit E (2007) Self-assembled peptide nanostructures: the design of molecular building blocks and their technological utilization. Chem Soc Rev 36:1263-1269

Gu Q, Cheng C, Gonela R, Suryanarayanan S, Anabathula S, Dai K, Haynie DT (2006) DNA nanowire fabrication. Nanotechnology 17:R14-R25

Hall SR, Bolger H, Mann S (2003) Morphosynthesis of complex inorganic forms using pollen grain templates. Chem Commun: 2784-2785

Hall SR, Shenton W, Engelhardt H, Mann S (2001) Site-specific organization of gold nanoparticles by biomolecular templating. Chem Phys Chem 2:184-186

Harnack O, Ford WE, Yasuda A, Wessels JM (2002) Tris(hydroxymethyl) phosphine-capped gold particles templated by DNA as nanowire precursors. Nano Lett 2:919-923

Harrison PM, Andrews SC, Artymuik PJ, Ford GC, Guest JR, Hirzmann J, Lawson DM, Livingstone JC, Smith JMA, Treffery A, Yewdall SJ, (1991) Probing structure-function relations in ferritin and bacterioferritin. Adv Inorg Chem 36: 449-486

Hasegawa T, Haraguchi S, Numata M, Fujisawa T, Li C, Kaneko K, Sakurai K, Shinkai S (2005) Schizophyllan can act as a one-dimensional host to construct poly (diacetylene) nanofibers. Chem Lett 34: 40-41

Henrick K, Thornton JM (1998) PQS: a protein quaternary structure file server. Trends Biochem Sci 23: 358-361

Hess H, Clemmens J, Qin D, Howard J, Vogel V (2001) Light-controlled molecular shuttles made from motor proteins carrying cargo on engineered surfaces. Nano Lett 1: 235-239

Hincke MT, Gautron J, Panheleux M, Garcia-Ruiz J, McKee MD, Nys Y (2000) Identification and localization of lysozyme as a component of eggshell membranes and eggshell matrix. Matrix Biol 19: 443-453

Huang Y, Chiang C-Y, Lee SK, Gao Y, Hu EL, Yoreo JD, Belcher AM (2005) Programmable assembly of nanoarchitectures using genetically engineered viruses. Nano Lett 5: 1429-1434

Hyodo T, Sasahara K, Shimizu Y, Egashira M (2005) Preparation of macroporous SnO_2 films using PMMA microspheres and their sensing properties to NO_x and H_2. Sens Actuators B 106: 580-590

Imai H, Matsuta M, Shimizu K, Hirashima H, Negishi N (2000) Preparation of TiO_2 fibers with well-organized structures. J Mater Chem 10: 2005-2006

Iwahori K, Yoshizawa K, Muraoka M, Yamashita I (2005) Fabrication of ZnSe nanoparticles in the apoferritin cavity by designing a slow chemical reaction system.

Inorg Chem 44: 6393-6400
Ji Q, Shimizu T (2005) Chemical synthesis of transition metal oxide nanotubes in water using an iced lipid nanotube as a template. Chem Commun: 4411-4413
Kawahashi N, Matijević E (1991) Preparation of hollow spherical particles of yttrium compounds. J Colloid Interface Sci 143:103-110
Kim Y (2003) Small structures fabricated using ash-forming biological materials as templates. Biomacromolecules 4: 908-913
Kim Y, Jung E (2002) Fabrication of discrete materials by interface-selective sol–gel polymerization. Chem Lett 31: 992-993
Kinsella JM, Ivanisevic A (2005) Enzymatic clipping of DNA wires coated with magnetic nanoparticles. J Am Chem Soc 127: 3276-3277
Kinsella JM, Ivanisevic A (2007) DNA-templated magnetic nanowires with different compositions: fabrication and analysis. Langmuir 23: 3886-3890
Klem MT, Resnick DA, Gilmore K, Young M, Idzerda YU, Douglas T (2007) Synthetic control over magnetic moment and exchange bias in all-oxide materials encapsulated within a spherical protein cage. J Am Chem Soc 129: 197-201
Knez M, Bittner AM, Boes F, Wege C, Jeske H, Mai E, Kern K (2003) Biotemplate synthesis of 3-nm Nickel and Cobalt nanowires. Nano Lett 3: 1079-1082
Knez M, Kadri A, Wege C, Gösele U, Jeske H, Nielsch K (2006) Atomic layer deposition on biological macromolecules: metal oxide coating of tobacco mosaic virus and ferritin. Nano Lett 6: 1172-1177
Knez M, Sumser MP, Bittner AM, Wege C, Jeske H, Hoffmann DMP, Kuhnke K, Kern K, (2004a) Binding the tobacco mosaic virus to inorganic surfaces. Langmuir 20: 441-447
Knez M, Sumser MP, Bittner AM, Wege C, Jeske H, Martin TP, Kern K (2004b) Spatially selective nucleation of metal clusters on the tobacco mosaic virus. Adv Funct Mater 14: 116-124
Kong J, Franklin NR, Zhou CW, Chapline MG, Peng S, Cho KJ, Dai HJ (2000) Nanotube molecular wires as chemical sensors. Science 287: 622-625
Koumoto K, Kimura T, Mizu M, Kunitake T, Sakurai K, Shinkai S (2002) Polysaccharide-polynucleotide complexes. Part 12. Enhanced affinity for various polynucleotide chains by site-specific chemical modification of schizophyllan. J Chem Soc Perkin Trans 1: 2477-2484
Kramer RM, Li C, Carter DC, Stone MO, Naik RR (2004) Engineered protein cages for nanomaterial synthesis. J Am Chem Soc 126: 13282-13286
Kumar A, Pattarkine M, Bhadbhade M, Mandale AB, Ganesh KN, Datar SS, Dharmadhikari CV, Sastry M (2001) Linear superclusters of colloidal gold particles by electrostatic assembly on DNA templates. Adv Mater 13: 341-344
Kumara MT, Tripp BC, Muralidharan S (2007a) Exciton energy transfer in self-assembled quantum dots on bioengineered bacterial flagella nanotubes. J Phys Chem C 111: 5276-5280
Kumara MT, Tripp BC, Muralidharan S (2007b) Self-assembly of metal nanoparticles and nanotubes on bioengineered flagella scaffolds. Chem Mater 19: 2056-2064
Li C, He J (2006) Easy replication of pueraria lobata toward hierarchically ordered porous γ-Al_2O_3. Langmuir 22: 2827-2831
Li C, Numata M, Bae A-H, Sakurai S, Shinkai S (2005) Self-assembly of supramolecular chiral insulated molecular wire. J Am Chem Soc 127: 4548-4549

Li M, Viravaidya C, Mann S (2007) Polymer-mediated synthesis of ferritin-encapsulated inorganic nanoparticles. Small 3: 1477-1481

Li S, Zeng Q, Xiao Y, Fu S, Zhou B (1995) Biomimicry of bamboo bast fiber with engineering composite materials. Mater Sci Eng C 3: 125-130

Li X, Fan T, Zhou H, Chow S, Zhang W, Zhang D, Guo Q, Ogawa H (2009) Enhanced light-harvesting and photocatalytic properties in morph-TiO_2 from green-leaf biotemplates. Adv Funct Mater 19: 45-46

Li X, Li Y, Cai J, Zhang D (2003) Metallization of bacteria cells. Sci China Ser E 46:161-167

Liu H, Freund MS (1997) New approach for the controlled cross-linking of polyaniline: synthesis and characterization. Macromolecules 30: 5660-5665

Losic D, Mitchell JG, Voelcker NH (2006) Fabrication of gold nanostructures by templating from porous diatom frustules. New J Chem 30: 908-914

Lu Y, Yin Y, Xia Y (2001) Preparation and characterization of micrometer-sized “egg shells”. Adv Mater 13: 271-274

Lvov YM, Price RR, Selinger JV, Singh A, Spector MS, Schnur JM (2000) Imaging nanoscale patterns on biologically derived microstructures. Langmuir 16: 5932-5935

Ma Y, Zhang J, Zhang G, He H (2004) Polyaniline nanowires on Si surfaces fabricated with DNA templates. J Am Chem Soc 126: 7097-7101

Mackle P, Charnock JM, Garner CD, Meldrum FC, Mann S (1993) Characterization of the manganese core of reconstituted ferritin by X-ray absorption spectroscopy. J Am Chem Soc 115: 8471-8472

Maddocks AR, Harris AT (2009) Biotemplated synthesis of novel porous SiC. Mater Lett 63: 748-750

Mahltig B, Haufe H, Böttcher H (2005) Functionalisation of textiles by inorganic sol–gel coatings. J Mater Chem 15: 4385-4398

Mallick D, Chakrabarti O, Bhattacharya D, Mukherjee M, Maiti HS, Majumdar R (2007) Electrical conductivity of cellular Si/SiC ceramic composites prepared from plant precursors. J Appl Phys 101: 033707

Mann S (1991) Flattery by imitation. Nature 349: 285-286

Mann S, Meldrum FC (1991) Controlled synthesis of inorganic materials using supramolecular assemblies. Adv Mater 3: 316-318

Mao C, Flynn CE, Hayhurst A, Sweeney R, Qi J, Georgiou G, Iverson B, Belcher AM (2003) Viral assembly of oriented quantum dot nanowires. Proc Natl Acad Sci USA 100: 6946-6951

Mao C, Solis DJ, Reiss BD, Kottmann ST, Sweeney RY, Hayhurst A, Georgiou G, Iverson B, Belcher AM (2004) Virus-based toolkit for the directed synthesis of magnetic and semiconducting nanowires. Science 303: 213-217

Mark SS, Bergkvist M, Bhatnagar P, Welch C, Goodyear AL, Yang X, Angert ER, Batt CA (2007) Thin film processing using S-layer proteins: biotemplated assembly of colloidal gold etch masks for fabrication of silicon nanopillar arrays. Colloids Surf B 57: 161-173

Mark SS, Bergkvist M, Yang X, Angert ER, Batt CA (2006a) Self-assembly of dendrimer-encapsulated nanoparticle arrays using 2-D microbial S-layer protein biotemplates. Biomacromolecules 7: 1884-1897

Mark SS, Bergkvist M, Yang X, Teixeira LM, Bhatnagar P, Angert ER, Batt CA (2006b) Bionanofabrication of metallic and semiconductor nanoparticle arrays using S-layer protein lattices with different lateral spacings and geometries. Langmuir 22: 3763-3774

Martín-Palma RJ, Pantano CG, Lakhtakia A (2008) Replication of fly eyes by the conformal-evaporated-film-by-rotation technique. Nanotechnology 19: 355704

McMillan RAHJ, Zaluzec NJ, Kagawa HK, Mogul R, Li YF, Paavola CD, Trent JD (2005) A self-assembling protein template for constrained synthesis and patterning of nanoparticle arrays. J Am Chem Soc 127: 2800-2801

Meldrum FC, Douglas T, Levi S, Arosio P, Mann S (1995) Reconstitution of manganese oxide cores in horse spleen and recombinant ferritins. J Inorg Biochem 58: 59-68

Mertig M, Ciacchi LC, Seidel R, Pompe W, Vita AD (2002) DNA as a selective metallization template. Nano Lett 2: 841-844

Mertig M, Kirsch R, Pompe W (1998) Biomolecular approach to nanotube fabrication. Appl Phys A 66: S723-S727

Mertig M, Wahl R, Lehmann M, Simon P, Pompe W (2001) Formation and manipulation of regular metallic nanoparticle arrays on bacterial surface layers: an advanced TEM study. Eur Phys J D 16: 317-320

Mizutani M, Takase H, Adachi N, Ota T, Daimon K, Hikichi Y (2005) Porous ceramics prepared by mimicking silicified wood. Sci Technol Adv Mater 6: 76-83

Monson CF, Woolley AT (2003) DNA-templated construction of copper nanowires. Nano Lett 3: 359-363

Morley KS, Marr PC, Webb PB, Berry AR, Allison FJ, Moldovan G, Brown PD, Howdle SM (2002) Clean preparation of nanoparticulate metals in porous supports: a supercritical route. J Mater Chem 12: 1898-1905

Muthukrishnan G, Hutchins BM, Williams ME, Hancock WO (2006) Transport of semiconductor nanocrystals by kinesin molecular motors. Small 2: 626-630

Nagarajan R, Liu W, Kumar J, Tripathy SK, Bruno FF, Samuelson LA (2001) Manipulating DNA conformation using intertwined conducting polymer chains. Macromolecules 34: 3921-3927

Nakao H, Shiigi H, Yamamoto Y, Tokonami S, Nagaoka T, Sugiyama S, Ohtani T (2003) Highly ordered assemblies of Au nanoparticles organized on DNA. Nano Lett 3: 1391-1394

Nam KT, Kim DW, Yoo PJ, Chiang CY, Meethong N, Hammond PT, Chiang YM, Belcher AM (2006) Virus-enabled synthesis and assembly of nanowires for lithium ion battery electrodes. Science 312: 885-888

Namba K, Pattanayek R, Stubbs GJ (1989) Visualization of protein-nucleic acid interactions in a virus: Refined structure of intact tobacco mosaic virus at 2.9 Å resolution by X-ray fiber diffraction. J Mol Biol 208: 307-325

Norris DJ (2007) Materials science: silicon life forms. Nature 446: 146-147

Numata M, Asai M, Kaneko K, Bae AH, Hasegawa T, Sakurai K, Shinkai S (2005a) Inclusion of cut and as-grown single-walled carbon nanotubes in the helical superstructure of schizophyllan and curdlan (β-1,3-glucans). J Am Chem Soc 127: 5875-5884

Numata M, Asai M, Kaneko K, Hasegawa T, Fujita N, Kitada Y, Sakurai K, Shikai S, (2004a) Curdlan and Schizophyllan (β-1,3-glucans) can entrap single-wall carbon nanotubes in their helical superstructure. Chem Lett 33: 232-233

Numata M, Hasegawa T, Fujisawa T, Sakurai K, Shinkai S (2004b) β-1,3-Glucan (Schizophyllan) can act as a one-dimensional host for creation of novel poly(aniline) nanofiber structures. Org Lett 6: 4447-4450

Numata M, Li C, Bae A-H, Kaneko K, Sakurai K, Shinkai S (2005b) β-1,3-Glucan polysaccharide can act as a one-dimensional host to create novel silica nanofiber structures. Chem Commun: 4655-4657

Numata M, Sugiyasu K, Hasegawa T, Shinkai S (2004c) Sol–gel reaction using DNA as a template: an attempt toward transcription of DNA into inorganic materials. Angew Chem Int Ed 43: 3279-3283

Ogasawara W, Shenton W, Davis SA, Mann S (2000) Template mineralization of ordered macroporous Chitin-silica composites using a cuttlebone-derived organic matrix. Chem Mater 12: 2835-2837

Okuda M, Iwahori K, Yamashita I, Yoshimura H (2003) Fabrication of nickel and chromium nanoparticles using the protein cage of apoferritin. Biotechnol Bioeng 84: 187-194

Okuda M, Kobayashi Y, Suzuki K, Sonoda K, Kondoh T, Wagawa A, Kondo A, Yoshimura H (2005) Self-organized inorganic nanoparticle arrays on protein lattices. Nano Lett 5: 991-993

Ono Y, Kanekyo Y, Inoue K, Hojo J, Nango M, Shinkai S (1999) Preparation of novel hollow fiber silica using collagen fibers as a template. Chem Lett 28: 475-476

Ono Y, Nakashima K, Sano M, Kanekyo Y, Inoue K, Hojo J, Shinkai S (1998) Organic gels are useful as a template for the preparation of hollow fiber silica. Chem Commun: 1477-1478

Panhorst M, Brückl H, Kiefer B, Reiss G, Santarius U, Guckenberger R (2001) Formation of metallic surface structures by ion etching using a S-layer template. J Vac Sci Technol B 19: 722-724

Parker AR, Townley HE (2007) Biomimetics of photonic nanostructures. Nat Nanotechnol 2: 347-353

Patolsky F, Weizmann Y, Lioubashevski O, Willner I (2002) Au-nanoparticle nanowires based on DNA and polylysine templates. Angew Chem Int Ed 41: 2323-2327

Patolsky F, Weizmann Y, Willner I (2004) Actin-based metallic nanowires as bio-nanotransporters. Nat Mater 3: 692-695

Pattanayek R, Stubbs GJ (1992) Structure of the U2 strain of tobacco mosaic virus refined at 3.5 Å resolution using X-ray fiber diffraction. J Mol Biol 228: 516-528

Payne EK, Rosi NL, Xue C, Mirkin CA (2005) Sacrificial biological templates for the formation of nanostructured metallic microshells. Angew Chem Int Ed 44: 5064-5067

Pérez-Cabero M, Puchol V, Beltrán D, Amorós P (2008) Thalassiosira pseudonana diatom as biotemplate to produce a macroporous ordered carbon-rich material. Carbon 46: 297-304

Pouget E, Dujardin E, Cavalier A, Moreac A, Valéry C, March-artzner V, Weiss T, Renault A, Paternostre M, Artzner F (2007) Hierarchical architectures by synergy between dynamical template self-assembly and biomineralization. Nat Mater 6: 434-439

Pum D, Schuster B, Sára M, Sleytr UB (2004) Functionalisation of surfaces with S-layers. IEEE Proc Nanobiotechnology 151: 83-86

Qi P, Vermesh O, Grecu M, Javey A, Wang O, Dai HJ, Peng S, Cho KJ (2003) Toward large arrays of multiplex functionalized carbon nanotube sensors for highly sensitive and selective molecular detection. Nano Lett 3: 347-351

Radloff C, Vaia RA, Brunton J, Bouwer GT, Ward VK (2005) Metal nanoshell assembly on a virus bioscaffold. Nano Lett 5: 1187-1191

Rambo CR, Sieber H (2005) Novel synthetic route to biomorphic Al_2O_3 ceramics. Adv Mater 17: 1088-1091

Reches M, Gazit E (2003) Casting metal nanowires within discrete self-assembled peptide nanotubes. Science 300: 625-627

Resnick DA, Gilmore K, Idzerda YU, Klem MT, Allen M, Douglas T, Young M, Arenholz E (2006) Magnetic properties of Co_3O_4 nanoparticles mineralized in Listeria innocua Dps. J Appl Phys 99: 08Q501

Richter J, Mertig M, Pompe W, Mönch I, Schackert HK (2001) Construction of highly conductive nanowires on a DNA template. Appl Phys Lett 78: 536-538

Richter J, Seidel R, Kirsch R, Mertig M, Pompe W, Plaschke J, Schackert HK (2000) Nanoscale palladium metallization of DNA. Adv Mater 12: 507-510

Rosi NL, Thaxton CS, Mirkin CA (2004) Control of nanoparticle assembly by using DNA-modified diatom templates. Angew Chem Int Ed 43: 5500-5503

Sakintuna B, Yurum Y, (2005) Templated porous carbons: a review article. Ind Eng Chem Res 44: 2893-2902

Sanchez C, Arribart H, Guille MMG (2005) Biomimetism and bioinspiration as tools for the design of innovative materials and systems. Nat Mater 4: 277-288

Sára M, Pum D, Schuster B, Sleytr UB (2005) S-Layers as patterning elements for application in nanobiotechnology. J Nanosci Nanotechnol 5: 1939-1953

Sastry M, Kumar A, Datar S, Dharmadhikari CV, Ganesh KN (2001) DNA-mediated electrostatic assembly of gold nanoparticles into linear arrays by a simple drop-coating procedure. Appl Phys Lett 78: 2943-2945

Scheibel T (2005) Protein fibers as performance proteins: new technologies and applications. Curr Opin Biotechnol 16: 427-433

Selvakannan PR, Swami A, Srisathiyanarayanan D, Shirude PS, Pasricha R, Mandale AB, Sastry M (2004) Synthesis of aqueous Au core-Ag shell nanoparticles using tyrosine as a pH-dependent reducing agent and assembling phase-transferred silver nanoparticles at the air–water interface. Langmuir 20: 7825-7836

Shemer G, Krichevski O, Markovich G, Molotsky T, Lubovitz I, Kotlyar AB (2006) Chirality of silver nanoparticles synthesized on DNA. J Am Chem Soc 128: 11006-11007

Shenton W, Douglas T, Young M, Stubbs G, Mann S (1999) Inorganic-organic nanotube composites from template mineralization of tobacco mosaic virus. Adv Mater 11: 253-256

Shenton W, Pum D, Sleytr UB, Mann S (1997) Synthesis of cadmium sulphide superlattices using self-assembled bacterial S-layers. Nature 389: 585-587

Shim M, Javey A, Kam NWS, Dai HJ (2001) Polymer functionalization for air-stable n-type carbon nanotube field-effect transistors. J Am Chem Soc 123: 11512-11513

Shin Y, Liu J, Chang JH, Nie Z, Exarhos GJ (2001) Hierarchically ordered ceramics through surfactant-templated sol–gel mineralization of biological cellular structures. Adv Mater 13: 728-732

Sleytr UB, Gyorvary E, Pum D (2003) Crystallization of S-layer protein lattices on surfaces and interfaces. Prog Org Coat 47: 279-287

Sleytr UB, Messner P, Pum D, Sára M (1999) Crystalline bacterial cell surface layers (S-layers): from supramolecular cell structure to biomimetics and nanotechnology.

Angew Chem Int Ed 38: 1034-1054
Sotiropoulou S, Sierra-sastre Y, Mark SS, Batt CA (2008) Biotemplated nanostructured materials. Chem Mater 20: 821-834
Spanhel L, Anderson MA (1991) Solvent and secondary kinetic isotope effects for the microhydrated SN2 reaction of Cl-$(H_2O)_n$ with CH_3Cl. J Am Chem Soc 113: 826-832
Stavenga DD (2002) Colour in the eyes of insects. J Comp Physiol A 188: 337-348
Stsiapura V, Sukhanova A, Baranov A, Artemyev M, Kulakovich O, Oleinikov V, Pluot M, Cohen JHM, Nabiev I (2006) DNA-assisted formation of quasi-nanowires from fluorescent CdSe/ZnS nanocrystals. Nanotechnology 17: 581-587
Suh DJ, Park T (2002) Synthesis of high-surface-area zirconia aerogels with a well-developed mesoporous texture using CO_2 supercritical drying. Chem Mater 14: 1452-1454
Sumper M, Brunner E (2006) Learning from diatoms: nature's tools for the production of nanostructured silica. Adv Funct Mater 16:17-26
Tans SJ, Devoret MH, Dai HJ, Thess A, Smalley RE, Geerligs LJ, Dekker C (1997) Individual single-wall carbon nanotubes as quantum wires. Nature 386: 474-477
Ueno T, Suzuki M, Goto T, Matsumoto T, Nagayama K, Watanabe Y (2004) Size-selective olefin hydrogenation by a Pd nanocluster provided in an apo-ferritin cage. Angew Chem Int Ed 43: 2527-2530
Unocic RR, Zalar FM, Sarosi PM, Cai Y, Sandhage KH (2004) Anatase assemblies from algae: coupling biological self-assembly of 3-D nanoparticle structures with synthetic reaction chemistry. Chem Commun: 796-797
Valentini L, Armentano I, Kenny JM, Cantalini C, Lozzi L, Santucci S (2003) Sensors for sub-ppm NO_2 gas detection based on carbon nanotube thin films. Appl Phys Lett 82: 961-963
Vallet-Regí M, Nicolopoulos S, Román J, Martínez JL, González-Calbet JM (1997) Structural characterization of ZrO_2 nanoparticles obtained by aerosol pyrolysis. J Mater Chem 7: 1017-1022
van Bommel KJC, Friggeri A, Shinkai S (2003) Organic templates for the generation of inorganic materials. Angew Chem Int Ed 42: 980-999
van den Heuvel MGL, Butcher CT, Smeets RMM, Diez S, Dekker C (2005) High rectifying efficiencies of microtubule motility on Kinesin-coated gold nanostructures. Nano Lett 5: 1117-1122
Varghese OK, Kichambre PD, Gong D, Ong KG, Dickey EC, Grimes CA (2001) Gas sensing characteristics of multi-wall carbon nanotubes. Sens Actuators B 81: 32-41
Verghese MM, Ramanathan K, Ashraf SM, Kamalasanan MN, Malhotra BD (1996) Electrochemical growth of polyaniline in porous sol–gel films. Chem Mater 8: 822-824
Vincenta JFV, Wegst U G K (2004) Design and mechanical properties of insect cuticle. Arthropod Struct Dev 33: 187-199
Wahl R, Mertig M, Raff J, Selenska-Pobell S, Pompe W (2001) Electron-beam induced formation of highly ordered palladium and platinum nanoparticle arrays on the S-layer of *Bacillus sphaericus* NCTC 9602. Adv Mater 13: 736-740
Wakayama H, Itahara H, Tatsuda N, Inagaki S, Fukushima Y (2001) Nanoporous metal oxides synthesized by the nanoscale casting process using supercritical fluids. Chem Mater 13: 2392-2396

Wang J, He S, Xie S, Xu L, Gu N (2007) Probing nanomechanical properties of nickel coated bacteria by nanoindentation. Mater Lett 61: 917-920

Wang Y, Liu Z, Han B, Huang Y, Yang G (2005a) Carbon microspheres with supported silver nanoparticles prepared from pollen grains. Langmuir 21: 10846-10849

Wang Y, Liu Z, Han B, Sun Z, Du J, Zhang J, Jiang T, Wu W, Miao Z (2005b) Replication of biological organizations through a supercritical fluid route. Chem Commun: 2948-2950

Wang Y, Lu L, Zheng Y, Chen X (2006) Improvement in hydrophilicity of PHBV films by plasma treatment. J Biomed Mater Res A 76A: 589-595

Warne B, Kasyutich OI, Mayes EL, Wiggins JAL, Wong KKW (2000) Self assembled nanoparticulate Co:Pt for data storage applications. IEEE Trans Magn 36: 3009-3011

Weatherspoon MR, Dickerson MB, Wang G, Cai Y, Shian S, Jones SC, Marder SR, Sandhage KH (2007) Thin, conformal, and continuous SnO_2 coatings on three-dimensional biosilica templates through hydroxy-group amplification and layer-by-layer alkoxide deposition. Angew Chem Int Ed 46: 5724-5727

Winningham TA, Gillis HP, Chouto DA, Martin KP, Moore JT, Douglas K (1998) Formation of ordered nanocluster arrays by self-assembly on nanopatterned Si(100) surfaces. Surf Sci 406: 221-228

Winningham TA, Whipple SG, Douglas K, (2001) Pattern transfer from a biomolecular nanomask to a substrate via an intermediate transfer layer. J Vac Sci Technol B 19: 1796-1802

Wong KKW, Mann S (1996) Biomimetic synthesis of cadmium sulfide-ferritin nano-composites. Adv Mater 8: 928-932

Wong M, Hendrix MJC, VonderMark K, Little C, Stern R (1984) Collagen in the egg shell membranes of the hen. Dev Biol 104: 28-36

Xie G, Zhang G, Lin F, Zhang J, Liu Z, Mu S (2008) The fabrication of subwavelength anti-reflective nanostructures using a bio-template. Nanotechnology 19: 095605

Xin H, Woolley AT (2003) DNA-templated nanotube localization. J Am Chem Soc 125:8710

Yamashita I, Hayashi J, Hara M (2004) Bio-template synthesis of uniform CdSe nanoparticles using cage-shaped protein, apoferritin. Chem Lett 33: 1158-1159

Yan H, Park SH, Finkelstein G, Reif JH, LaBean TH (2003) DNA-templated self-assembly of protein arrays and highly conductive nanowires. Science 301: 1882-1884

Yang D, Qi L, Ma J (2002) Eggshell membrane templating of hierarchically ordered macroporous networks composed of TiO_2 tubes. Adv Mater 14: 1543-1546

Zaitlin M (2000) Tobacco mosaic virus. AAB Descriptions of Plant Viruses 370: 8

Zhang B, Davis SA, Mendelson NH, Mann S (2000) Bacterial templating of zeolite fibres with hierarchical structure. Chem Commun: 781-782

Zhang D, Qi L (2005) Synthesis of mesoporous titania networks consisting of anatase nanowires by templating of bacterial cellulose membranes. Chem Commun: 2735-2737

Zhang G, Yang J, Ohji T (2001) Fabrication of porous ceramics with unidirectionally aligned continuous pores. J Am Ceram Soc 84: 1395-1397

Zhang J, Liu Y, Ke Y, Yan H (2006a) Periodic square-like gold nanoparticle arrays templated by self-assembled 2D DNA nanogrids on a surface. Nano Lett 6: 248-251

Zhang W, Zhang D, Fan T, Ding J, Guo Q, Ogawa H (2006b) Fabrication of ZnO microtubes with adjustable nanopores on the walls by the templating of butterfly wing scales. Nanotechnology 17: 840-844

Zhang Z, Buitenhuis J (2007) Synthesis of uniform silica rods, curved silica wires, and silica bundles using filamentous fd virus as a template. Small 3: 424-428

Zhou H, Fan T, Zhang D (2007) Hydrothermal synthesis of ZnO hollow spheres using spherobacterium as biotemplates. Micropor Mesopor Mater 100: 322-327

3

Inorganic-organic Hybrid Materials Based on Nano-polyoxometalates and Surfactants

Peiqin Tang, Jingcheng Hao*
Key Laboratory for Colloid and Interface Chemistry, Ministry of Education, Shandong University, Jinan, 250100, China. *Email: jhao@sdu.edu.cn

A series of nano-scale polyoxometalates (POMs) with beautiful topologies have been successfully synthesized by destroying the hydration shell of metal oxides. Their magnetic, electronic, and photoluminescent properties and valuable applications in catalysis, medicine, and material science are discussed. By interaction with oppositely charged surfactants, an interesting phase transfer of POMs occurs from aqueous solution to chloroform solution by forming hydrophobic surfactant-encapsulated clusters (HSECs), and then to aqueous solution with more surfactants. Moreover, inorganic-organic hybrids composed of polyoxometalates and surfactants or polyelectrolytes are fabricated by layer-by-layer (LbL) and Langmuir-Blodgett (LB) techniques. A simple solvent-evaporation method is also delicately carried out, and consequently, ordered honeycomb films are self-assembled at air/water interfaces without any extra moist airflow. It is speculated that the condensed water micro-droplets induced by the quick evaporation of solvents play an important role as template for the formation of pores. And many types of modulation are being investigated to construct thin films with different morphologies. The incorporation of functional polyoxometalates in thin films will bridge polyoxometalate chemistry and material chemistry.

3.1 Introduction to Developed POMs

Early transition metals (M = Mo, V, W, Nb, Ta) have been known for over 200 years, since the time of Scheele in 1783 and Berzelius in 1826. In their highest oxidation state, they can form discrete nano-scale oxygen cluster anions which are named polyoxometalates (POMs). However, it was impossible to get pure crystalline compounds from the molybdenum blue solutions obtained by reducing an acidified molybdate solution, mainly due to the high solubility of the abundant cluster anions. In the last three decades, Professor Achim Müller succeeded in synthesizing various polyoxometalates by destroying the hydration shell of the anions caused by the extremely hydrophilic surface (Müller, Serain, 2000). Now, POMs are becoming a large and rapidly-growing class of compounds.

3.1.1 Structures of POMs

The structures of nano-scale POMs are based upon $\{MO_x\}$ polyhedra sharing vertices, edges, or more rarely, faces. A large variety of POMs can be obtained by linking metal-oxygen building units either as existing or virtual species. Compared with the traditional salts, POMs are composed of many more atoms (even thousands), and become giant, with diameters of several nanometers. Moreover, POMs reveal a huge variety of shapes, sizes and compositions, such as $\{Mo_{37}\}((NH_4)_{14}[H_{14}Mo_{37}O_{112}] \cdot 35H_2O)$ (Müller et al., 1998a), a "Keplerate-type" sphere $\{Mo_{132}\}$ $((NH_4)_{42}[H_{234}Mo_{132}O_{504}] \cdot 300\ H_2O \cdot 10\ CH_3COONH_4)$ (Müller et al., 1998b) and $\{Mo_{72}Fe_{30}\}$ $(Mo_{72}Fe_{30}O_{252}L_{102} \cdot x\ H_2O$ ($x \approx 180$, L=H_2O, CH_3COO^-, $Mo_2O^{n-}{}_{8/9}$)) (Müller et al., 1999a), a "giant wheel-type" cluster $\{Mo_{154}\}$ $(Na_{15}[H_{152}Mo_{154}O_{529}] \cdot$ *ca.* 400 H_2O) (Müller et al., 1999b), a "protein-sized blue lemon" $\{Mo_{368}\}$ $(Na_{48}[H_{496}Mo_{368}O_{1464}S_{48}] \cdot$ *ca.* 1000 H_2O) (Müller et al., 2002; 2004), and sandwich-type clusters, $\{Mn_2Bi_2W_{20}\}$ $(Na_6(NH_4)_4[(Mn(H_2O)_3)_2(WO_2)_2(BiW_9O_{33})_2])$ (Bösing et al., 1998) and $\{EuW_{10}\}$ $(Na_9[EuW_{10}O_{36}] \cdot 32\ H_2O)$ (Sugeta, Yamase, 1993). Examples of molybdenum-oxide based building blocks (Müller et al., 1999c) and the structures of POMs are schematically shown in Fig.3.1.

There are two generic families of POMs (Pope, Müller, 1991). One is the isopoly compounds (also called isopolyanions or isopolyoxometalates) that contain only the d^0 metal cations and oxide anions, such as $\{Mo_{37}\}$, $\{Mo_{132}\}$, $\{Mo_{154}\}$, and $\{Mo_{368}\}$. The other is the heteropoly compounds (also called heteropolyanions or hetero-polyoxometalates) that contain one or more p-, d-, or f-block "heteroatoms" in addition to other ions, such as $\{Mn_2Bi_2W_{20}\}$, $\{EuW_{10}\}$, and $\{PW_{12}\}(H_3PW_{12}O_{40})$ (Brown et al., 1977). Over half of the elements in the periodic table are known to function as heteroatoms in heteropoly compounds. These heteroatoms can reside either buried in (not solvent accessible) or at surface (solvent accessible) positions in the POM structure.

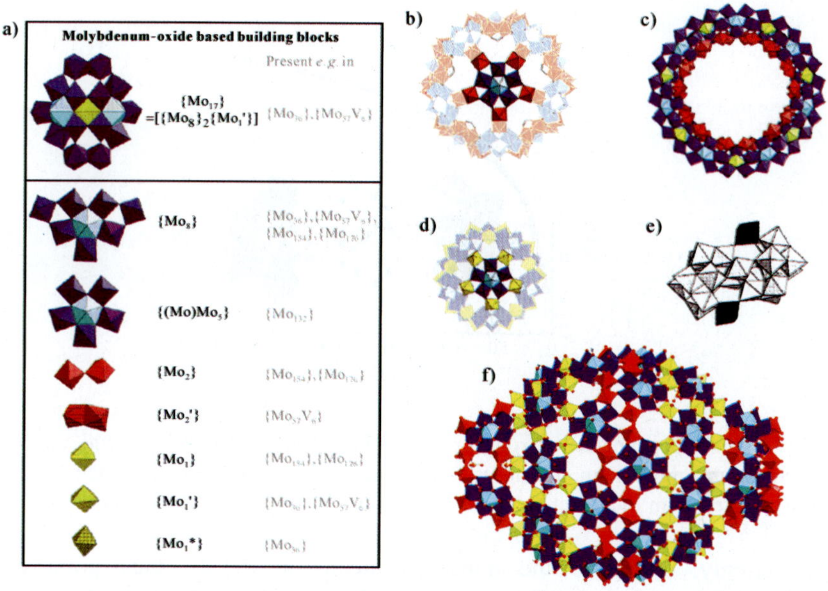

Fig.3.1. a) Schematic representation of some molybdenum-oxide based building blocks. Reprinted from Müller et al., 1999c. Copyright (1999), with permission from RSC. And the schematic structures: b) $\{Mo_{132}\}$; c) $\{Mo_{154}\}$; d) $\{Mo_{72}Fe_{30}\}$; e) $\{Mn_2Bi_2W_{20}\}$; and f) $\{Mo_{368}\}$. Reprinted from Müller et al., 1999a; 2002; Bösing et al., 1998. Copyright (1999, 2002, and 1998), with permission from Wiley, and ACS

3.1.2 Properties of POMs

The pioneers, Francis Secheresse, Henryk Ratajczak, and Achim Müller had expected that "Inorganic metal-oxygen cluster anions form a class of compounds that is unique in its topological and electronic versatility and is important in several disciplines" (Secheresse et al., 2002). Many reports indicate that nano-scale POM clusters are highly functional. The magnetic properties of the high-nuclearity Ni^{II}-substituted polyoxometalate $Na_{15}[Na\{(A\text{-}\alpha\text{-}SiW_9O_{34})Ni_4(CH_3COO)_3(OH)_3\}_2] \cdot 4NaCl \cdot 36H_2O$ ($\{Ni_4SiW_9\}$) were studied (Pichon et al., 2008). Measurements of magnetic susceptibility revealed the occurrence of concomitant ferromagnetic and antiferromagnetic interactions in $\{Ni_4SiW_9\}$ clusters. The extension of magnetic exchange has also been determined by means of a spin Hamiltonian (Fig.3.2). The magnetic properties of nano-size giant Keplerate species $\{Mo_{72}Fe_{30}\}$, which is a very large paramagnetic molecule, were also investigated by experimental investigation and theoretical simulation with the classical and quantum Heisenberg model (Müller et al., 2001). Recently, the magnetic properties of other discrete molecular transition metal- and lanthanide-containing polyoxotungstates, polyoxomolybdates, and polyoxonickelates have also been studied (Kortz et al., 1999; 2009; Prinz et al., 2006).

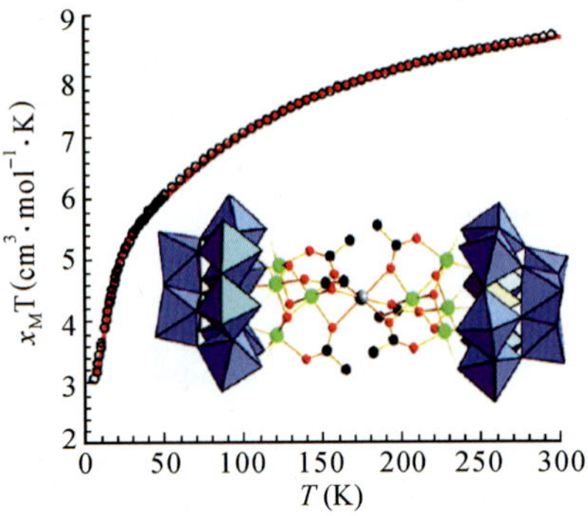

Fig.3.2. Thermal dependence of x_MT for $\{Ni_4SiW_9\}$. The solid line represents the best-fit parameters using Hamiltonian. The inset schematically shows the structure of the $\{Ni_4SiW_9\}$ cluster. Reprinted from Pichon et al., 2008. Copyright (2008), with permission from ACS

Interestingly, owing to the transition metals in POMs, they possess rich electrochemical properties (López et al., 2001; Jabbour et al., 2005; Keita et al., 2007). Taking Dawson-type tungstophosphates $[H_4PW_{18}O_{62}]^{7-}$(PW_{18}) and $[P_2W_{18}O_{62}]^{6-}$ (P_2W_{18}) clusters as examples for detailed description (Mbomekalle et al., 2004), the cyclic voltammograms (CVs) of the two complexes in a pH 0.3 sulfate medium are shown superimposed (Fig.3.3a). The first wave of PW_{18} unambiguously represents a two-electron chemically reversible process. This point was checked by controlled potential coulometry. The two subsequent waves also feature two-electron processes. In short, the pattern for PW_{18} is composed of a set of three reversible diffusion-controlled waves. The authors expanded this observation by a brief study of the pattern as a function of pH. This point is illustrated in Fig.3.3b, which shows, in superimposition, the CVs of PW_{18} at pH 0.3 and 4.0. With the increase in pH, the formerly two-electron wave splits into two apparently one-electron processes. In addition, in different acidic or basic conditions, the redox properties of such heteropolyanions (HPAs) can be changed a little. The voltammetric behavior of PW_{18} towards the reduction of nitrite suggests that the addition of NO_2^- induces a large current intensity enhancement (Fig.3.3c). No anodic trace is observed on potential reversal, even for small values of the excess parameter γ ($\gamma = C_{NO_2^-} / C_{PW_{18}}$). Increasing the concentration of nitrite still enhances the cathodic current. It must be concluded that an efficient catalysis of the reduction of nitrite can be achieved during the reduction of PW_{18}. Some applications of POMs in catalysis will be described in the following section. Some lanthanide-containing and transition metal-containing polyoxometalates are photoluminescent (Wang et al., 2006; Ballardini et al., 1984). For example, a Eu-containing $\{EuW_{10}\}^{9-}$ polyanion is strongly luminescent under UV light excitation (Fig.3.4).

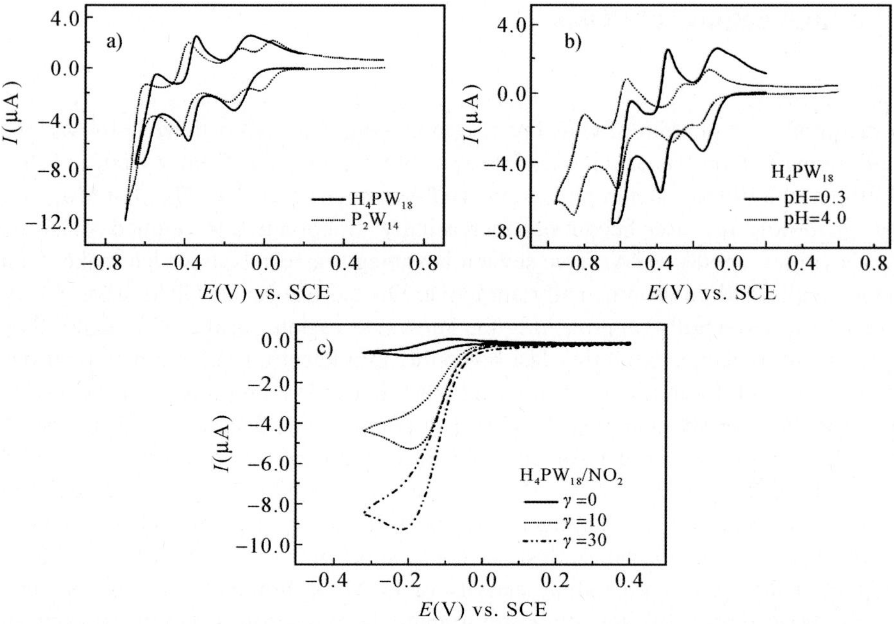

Fig.3.3. a) Comparison of the cyclic voltammograms (CVs) measured in a pH 0.3 medium (0.2 M Na_2SO_4 + H_2SO_4) for 5 × 10^{-4} M solutions of the two tungstophosphates: PW_{18} and P_2W_{18}. b) pH effects on the CVs of PW_{18}. c) CVs showing the catalytic activity of PW_{18} towards nitrite at pH 1 (0.2 M Na_2SO_4 + H_2SO_4) of 2 × 10^{-4} M PW_{18} in the presence of increasing amounts of nitrite NO_2^-, and the excess parameter $\gamma = C_{NO_2^-} / C_{PW_{18}}$ is indicated. CV conditions: working electrode: glassy carbon; reference electrode: SCE; scan rate: 10 mV·s^{-1} for a) and b), 2 mV·s^{-1} for c). Reprinted from Mbomekalle et al., 2004. Copyright (2004), with permission from Wiley

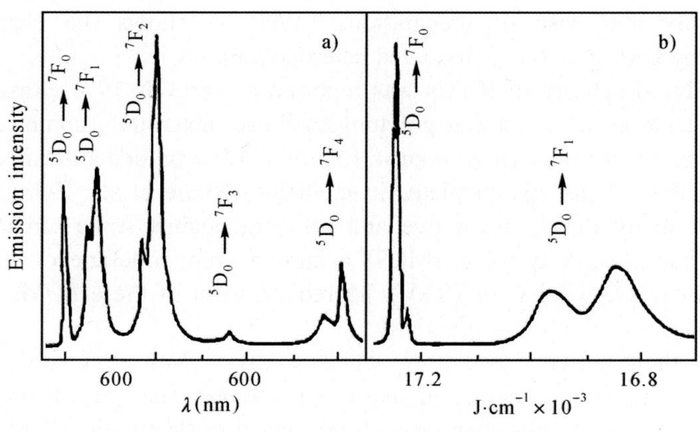

Fig.3.4. Luminescence spectra of 5.0 × 10^{-5} M $\{EuW_{10}\}^{9-}$ solutions at a) low (λ_{exc} = 270 nm) and b) high (λ_{exc} = 465.8 nm) resolutions. Reprinted from Ballardini et al., 1984. Copyright (1984), with permission from ACS

3.1.3 Applications of POMs

The intrinsic properties of polyoxometalates are of interest in themselves not only from a fundamental point of view but also in making materials of interest for various applications (Casan-Pastor, Gómez-Romero, 2004; Coronado, Gomez-Garcia, 1998; Müller et al., 1998c). Heteropolyanions (HPAs), such as $H_3PW_{12}O_{40}$, $H_3PMo_{12}O_{40}$, and $H_4SiMo_{12}O_{40}$, are becoming increasingly important for applied catalysis (Kozhevnikov, 1998). HPAs have several advantages as catalysts which make them economically and environmentally attractive. On the one hand, HPAs have a very strong Brönsted acidity, approaching the superacid region; on the other hand, they are efficient oxidants, exhibiting fast reversible multielectron redox transformations under rather mild conditions. Their acid-base and redox properties can be varied over a wide range by changing the chemical composition. Moreover, HPAs have a very high solubility in polar solvents and fairly high thermal stability in the solid state. These properties render HPAs potentially promising acid, redox, and bifunctional catalysts in homogeneous as well as in heterogeneous systems. However, some POM-based catalysts still have some drawbacks (Proust et al., 2008): (i) in the solid state, the catalytic activity of POMs is limited owing to their low specific surface area; (ii) in comparison to many homogeneous catalytic systems, an apparent weakness of POMs is the absence of pathways to direct the substrate towards the catalytically active center, which could be bypassed by a suitable modification of the POM surface.

In addition, much attention has been focused on the electrocatalytic behavior of HPAs (Sadakane, Steckhan, 1998). However, these workers found that the number of polyoxometalates which could be used as electrocatalysts was limited. Mostly, α-Keggin and Dawson-type heteropolyanions of phosphotungstate, silicontungstate, phosphomolybdate, and silicomolybdate; mixed-addenda heteropolyanions; and transition metal-substituted heteropolyanions are used as electrocatalysts because of their stability and ease of preparation. Table 3.1 shows the electrocatalytic reductions by homogeneously dissolved heteropolyanions.

The antiviral activity of POMs was reported as early as 1971 (Raynaud et al., 1971). Raynaud et al. noted that polytungstosilicate heteropoly compounds inhibit murine leukemia sarcoma virus *in vitro*. In 1998, Hill provided a detailed review of the applications of polyoxometalates in medicine (Rhule et al., 1998). Now, it is known that many POMs are active and efficient against some viruses, such as murine leukemia sarcoma virus (MLSV), human immunodeficiency virus (HIV), and respiratory syncytial virus (RSV). Moreover, most of these POMs show good inhibitory activity with low cytotoxicity.

Beyond their conventional hotspots as catalysts and virus inhibitors, polyoxometalates are also attractive in many applications, such as corrosion-resistant coatings, dyes/pigments/inks, membranes, and capacitors (Katsoulis, 1998) (Table 3.2). Polyoxometalates have relatively low toxicity compared to chromates, which are highly toxic and the most widely used inhibitors to combat corrosion of metal

Table 3.1. Electrocatalytic reductions by homogeneously dissolved heteropolyanions. Reprinted from Sadakane, Steckhan, 1998. Copyright (1998), with permission from ACS

Catalyst	Reducton	Condition
$SiW_{12}O_{40}^{4-}$	h.e.r.	acidic solution
	nitrite reduction	buffer
	O_2 reduction	triflate buffer (pH=2)
$SiMo_{12}O_{40}^{4-}$	chlorate reduction	50% (v/v) dioxane-water
	nitrite reduction	(0.5 M H_2SO_4)
$PMo_{12}O_{40}^{3-}$	chlorate reduction	50% (v/v) dioxane-water (0.5 M H_2SO_4)
$P_2W_{18}O_{62}^{6-}$	nitrite reduction	buffer
$XW_{11}O_{39}Fe^{III}(H_2O)^{n-}$ (X=P, As, Si, Ge)	nitrite reduction	buffer
$SiW_{11}O_{39}Fe^{III}(H_2O)^{5-}$	H_2O_2 reduction	buffer
$P_2W_{17}O_{61}Fe^{III}(H_2O)^{8-}$	H_2O_2 reduction	sulfate buffer (pH=3)
	nitrite reduction	acetate buffer (pH=5)
$PW_{11}O_{39}Ru^{III}(H_2O)^{4-}$	DMSO reduction h.e.r.	sulfate buffer (pH=2)
$K_{17}[Ln(AsW_{17}O_{61})_2]$ (Ln=La, Rr, Sm, Eu, Gd, Dy, Tm)	nitrite reduction	acetate buffer (pH=5.5)
$Nd(SiMo_7W_4O_{39})_{21}^{3-}$	bromate reduction	buffer

Table 3.2. Categories of applications for POMs excluding the conventional applications in catalysis and medicine. Reprinted from Katsoulis, 1998. Copyright (1998), with permission from ACS

No.	Categories
1	coatings
2	analytical chemistry
3	processing radioactive waste
4	separations
5	sorbents of gases
6	membranes
7	sensors
8	dyes/pigments
9	electrooptics
10	electrochemistry/electrodes
11	capacitors
12	dopants in nonconductive polymers
13	dopants in conductive polymers
14	dopants in sol–gel matrixes
15	cation exchangers
16	flammability control
17	bleaching of paper pulp
18	clinical analysis
19	food chemistry

surfaces. Furthermore, polyoxometalates can accept electrons without major changes of their structures and form insoluble salts with large cations. These properties make them attractive as oxidizing and film-forming corrosion inhibitors. The ability of POMs to form stable precipitates with cationic dyes has resulted in considerable patent activity from industries using pigments, dyes, and inks. The high ionic conductivity of POMs, their capacity to form a plethora of salts with virtually any cation, and their ability to undergo redox processes under many mild conditions also make polyoxometalates a good source for membrane-based devices and sensors. These membranes and sensors can be useful in selective electrodes, in gas detection apparatus, in solid-state electrochromic devices, and in liquid and solid electrolytic cells.

3.2 Inorganic-organic Hybrids of Polyoxometalates and Surfactants/Polyelectrolytes

Owing to their sizes, structures, and compositions, POMs are generally soluble in polar solvents, including water and ethanol. Mostly, POMs are ionized as giant anions (several nanometers) and countered by small cations, such as Na^+, NH_4^+, and H^+, in aqueous solutions. According to electrostatic interaction by an ion-exchange reaction, hydrophilic POMs anions can be easily surface-modified by oppositely charged species, such as surfactants (Volkmer et al., 2000; 2002; Liu, 2003; Kurth et al., 2000a), polymers (Liu et al., 2002a), proteins (Lee et al., 2005), and porphyrin (Bazzan et al., 2008) molecules. This surface modification can improve the stability and the solubility of encapsulated POMs in nonpolar solvents. Moreover, it can change the surface chemical properties of POMs in order to realize the integration of multifunctional POMs into thin films and ordered three-dimensional aggregations. Now, much attention is focused on the construction of these organic-inorganic hybrids by layer-by-layer (LbL) (Liu et al., 2002b; Jiang et al., 2003) and Langmuir-Blodgett (LB) techniques (Clemente-León et al., 1997a; 1997b; 1998a; Liu et al., 1999), and other self-assembly methods (Bu et al., 2002; Sun et al., 2006). Here, we discuss the interaction of POMs and oppositely charged surfactants, which finally produce hydrophobic surfactant-encapsulated clusters (HSECs), which are novel organic-inorganic capsules with the head groups of surfactants interacting with POMs and the external hydrocarbon chains.

3.2.1 Phase Behavior of Mixtures of POMs and Surfactants

Some polyoxomolybdates, such as $\{Mo_{132}\}$, $\{Mo_{154}\}$, $\{Mo_{368}\}$, and $\{Mo_{72}Fe_{30}\}$, which are synthesized by reported methods (Müller et al., 1998b; 1999a; 1999b; 2004), are here taken as examples, and the phase behavior of their mixtures and cationic surfactants were investigated in detail by our group (Tang et al., 2008; Fan, Hao, 2009a). These polyoxomolybdates are quite soluble in water, because they contain a large number of water ligands and are negatively charged in solution (similar to polyelectrolytes). $\{Mo_{132}\}^{42-}$ in solution is balanced by small NH_4^+ counterions. $\{Mo_{154}\}$ and $\{Mo_{368}\}$ are medium-strong acids, and the giant polyanions are balanced by Na^+ and H^+. It is necessary to note that aqueous solutions of $\{Mo_{132}\}$, $\{Mo_{154}\}$, and $\{Mo_{368}\}$ are not stable in an open environment because of the oxidation by O_2 from Mo^V to Mo^{VI}. So, freshly prepared aqueous solutions of polyoxometalates are used, and bubbling with pure nitrogen to expel oxygen is needed. The $\{Mo_{72}Fe_{30}\}$ cluster has a little difference and exists as almost neutral molecules in crystal form. While $\{Mo_{72}Fe_{30}\}$ molecules behave like a weak acid in the solution state, the water ligands attached to the Fe^{III} center tend to partially deprotonate, thus making $\{Mo_{72}Fe_{30}\}$ clusters slightly negatively charged (carrying several localized charges). The degree of deprotonation depends on the pH of the solution (Liu et al., 2006a), from almost 0 at pH 3.0 to 22 at pH 4.9. We assumed that these POMs could interact with the oppositely charged surfactants, and a phase transition would be observed by controlling the concentration of the surfactants in solution. The cationic surfactant tetradecyltrimethylammonium bromide (TTABr) was used. Stock aqueous solutions of 0.1 mg·mL^{-1} $\{Mo_{132}\}$ (brown) and $\{Mo_{72}Fe_{30}\}$ (yellow) were prepared and kept under a N_2 atmosphere. Then, various amounts of TTABr were added into the $\{Mo_{132}\}$ or $\{Mo_{72}Fe_{30}\}$ solutions. The concentration of TTABr was from 0 to 1.0 mg·mL^{-1}, and the sample solutions were kept at (25.0±0.1) °C under a N_2 atmosphere for more than 4 weeks. The two systems showed the same phase behavior (Fig.3.5).

Fig.3.5. Photographs of samples showing the phase behavior of 0.1 mg·mL^{-1} $\{Mo_{132}\}$ and $\{Mo_{72}Fe_{30}\}$ aqueous solutions induced by controlling TTABr concentration. Upper: $\{Mo_{132}\}$-TTABr system, c_{TTABr} = 0.0 (A), 0.008 (B), 0.032 (C), 0.048 (D), 0.12 (E), 0.15 (F), and 1.0 mg·mL^{-1} (G). Lower: $\{Mo_{72}Fe_{30}\}$-TTABr system, c_{TTABr} = 0.0 (a), 0.016 (b), 0.024 (c), 0.068 (d), 0.20 (e), 0.70 (f), and 1.0 mg·mL^{-1} (g). Reprinted from Fan, Hao, 2009a. Copyright (2009), with permission from Elsevier

One can clearly see that the phase transitions were induced by different amounts of surfactant. In the $\{Mo_{132}\}$-TTABr system, when the TTABr concentration reached about 0.048 mg·mL^{-1}, *i.e.*, n_{TTABr}:$n_{\{Mo_{132}\}}$ = 40:1, the solution became completely clear and colorless, and a brown precipitate formed at the bottom of the tube, presumably containing insoluble $\{Mo_{132}\}$-$(TTA)_{40}$ complexes. Similarly, in the $\{Mo_{72}Fe_{30}\}$-TTABr system, when the TTABr concentration was 0.068 mg·mL^{-1}, *i.e.*, n_{TTABr}:$n_{\{Mo_{72}Fe_{30}\}}$ = 90:1, the solution became completely clear and colorless, and slightly yellow floccules formed, presumably containing insoluble $\{Mo_{72}Fe_{30}\}$-$(TTA)_{90}$ complexes. Interestingly, with increasing the TTABr concentration, the brown $\{Mo_{132}\}$-$(TTA)_{40}$ precipitate and the slightly yellow $\{Mo_{72}Fe_{30}\}$-$(TTA)_{90}$ floccules gradually dissolved, and the red-brown solution of $\{Mo_{132}\}$-TTABr and the yellow solution of $\{Mo_{72}Fe_{30}\}$-TTABr returned. The two systems became uniform and stable single phases, forming $\{Mo_{132}\}$- or $\{Mo_{72}Fe_{30}\}$-based bilayer-like structures to make hydrophilic complexes again. In detail, when $c_{TTABr}<0.048$ mg·mL^{-1} for the $\{Mo_{132}\}$-TTABr system, and $c_{TTABr}<0.068$ mg·mL^{-1} for the $\{Mo_{72}Fe_{30}\}$-TTABr system, the solution of $\{Mo_{132}\}$-$(TTA)_n$ ($n<40$) or $\{Mo_{72}Fe_{30}\}$-$(TTA)_n$ ($n<90$) has negative charge properties and the POMs surfaces are incompletely encapsulated. When $c_{TTABr} = 0.048$ mg·mL^{-1} for the $\{Mo_{132}\}$-TTABr system and $c_{TTABr} = 0.068$ mg·mL^{-1} for the $\{Mo_{72}Fe_{30}\}$-TTABr system at the stoichiometric ratios of $r_{TTA^+/\{Mo_{132}\}}$ and $r_{TTA^+/\{Mo_{72}Fe_{30}\}}$ in aqueous solution, the $\{Mo_{132}\}$-$(TTA)_{40}$ precipitates or $\{Mo_{72}Fe_{30}\}$-$(TTA)_{90}$ floccules form, and the POMs surfaces are completely encapsulated. With excess amounts of TTABr, the free cationic TTA^+ cations interact with $\{Mo_{132}\}$-$(TTA)_{40}$ or $\{Mo_{72}Fe_{30}\}$-$(TTA)_{90}$ complexes due to the hydrophobic interaction between hydrocarbon chains. So the precipitates and floccules dissolve and form hydrophilic $\{Mo_{132}\}$-$(TTA)_n$ ($n>40$) or $\{Mo_{72}Fe_{30}\}$-$(TTA)_n$ ($n>90$) complexes with the hydrophilic head groups of surfactants outside, which make the complexes positively charged. Zeta potential data of the two systems with various concentrations of TTABr are shown in Fig.3.6.

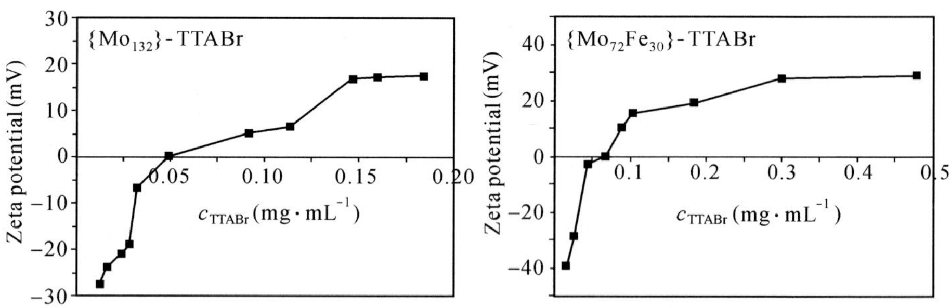

Fig.3.6. Zeta potential data of $\{Mo_{132}\}$- and $\{Mo_{72}Fe_{30}\}$-TTABr systems as a function of concentration of TTABr. The concentration of $\{Mo_{132}\}$ and $\{Mo_{72}Fe_{30}\}$ is 0.1 mg·mL^{-1}. Reprinted from Fan, Hao, 2009a. Copyright (2009), with permission from Elsevier

The size and shape of the $\{Mo_{132}\}$-$(TTA)_n$ and $\{Mo_{72}Fe_{30}\}$-$(TTA)_n$ complexes in solution were characterized by transmission electron microscopy (TEM). TEM images of $\{Mo_{132}\}$-$(TTA)_n$ complexes at c_{TTABr} = 0.028 (less than the stoichiometric ratio of $r_{TTA^+/\{Mo_{132}\}}$, n<40) and 0.18 mg·mL^{-1} TTABr (much more than the stoichiometric ratio of $r_{TTA^+/\{Mo_{132}\}}$, n>40) in 0.1 mg·mL^{-1} $\{Mo_{132}\}$ aqueous solutions are shown in Fig.3.7. When the concentration of TTABr is below 0.048 mg·mL^{-1}, *i.e.*, the precipitates and the $\{Mo_{132}\}$-$(TTA)_n$ (n<40) complexes coexist, the $\{Mo_{132}\}$-$(TTA)_n$ (n<40) complexes mainly exist as spheres. At c_{TTABr} = 0.18 mg·mL^{-1}, the $\{Mo_{132}\}$-$(TTA)_{40}$ precipitates dissolve and the brown isotropic solution returns. TEM observations showed that spherical $\{Mo_{132}\}$-$(TTA)_n$ (n>40) complexes still existed. In this case, the supposed $\{Mo_{132}\}$-$(TTA)_n$ (n>40) complexes at excess TTABr in solution should be $(TTA)_m[\{Mo_{132}\}$-$(TTA)_{40}]$ or $(TTA)_m\cdot[\{Mo_{132}\}$-$(TTA)_{40}]_x$ (x>1), which have hydrophilic surfaces.

TEM and high-resolution TEM (HR-TEM) images of $\{Mo_{72}Fe_{30}\}$-$(TTA)_n$ solutions with c_{TTABr} = 0.016 (Figs.3.8a and 3.8b), 0.1 (Fig.3.8c), and 0.3 mg·mL^{-1} (Fig.3.8d) in 0.1 mg·mL^{-1} $\{Mo_{72}Fe_{30}\}$ are shown in Fig.3.8. When c_{TTABr}<0.068 mg·mL^{-1}, *i.e.*, the stoichiometric ratio of $r_{TTA^+/\{Mo_{72}Fe_{30}\}}$, spherical $\{Mo_{72}Fe_{30}\}$-$(TTA)_n$ (n<90) complexes form, which were seen in TEM images (Figs.3.8a and 3.8b). $\{Mo_{72}Fe_{30}\}$ molecules in aqueous solution behave like a weak acid, making $\{Mo_{72}Fe_{30}\}$ clusters carry surface negative charges. The $\{Mo_{72}Fe_{30}\}$ macroanions can combine with few TTA^+ to form $\{Mo_{72}Fe_{30}\}$-$(TTA)_n$ complexes. When c_{TTABr}>0.068 mg·mL^{-1}, the floccules formed at the stoichiometric ratio of $r_{TTA^+/\{Mo_{72}Fe_{30}\}}$ start to gradually dissolve. Hydrophilic $\{Mo_{72}Fe_{30}\}$-$(TTA)_n$ (n>90) complexes form. When c_{TTABr}> 0.3 mg·mL^{-1}, the floccules dissolve completely and the system becomes a uniform and stable single-phase solution. The TEM images confirm the existence of spherical complexes, which are polydispersed. HR-TEM images (Fig.3.8d) reveal 2.5-nm-diameter objects attributed to $\{Mo_{72}Fe_{30}\}$ molecules. At excess TTABr, $\{Mo_{72}Fe_{30}\}$-$(TTA)_n$ (n>90), *i.e.*, the hydrophilic $(TTA)_m[\{Mo_{72}Fe_{30}\}$-$(TTA)_{90}]$ or $(TTA)_m\cdot[\{Mo_{72}Fe_{30}\}$-$(TTA)_{90}]_x$ (x>1) could form.

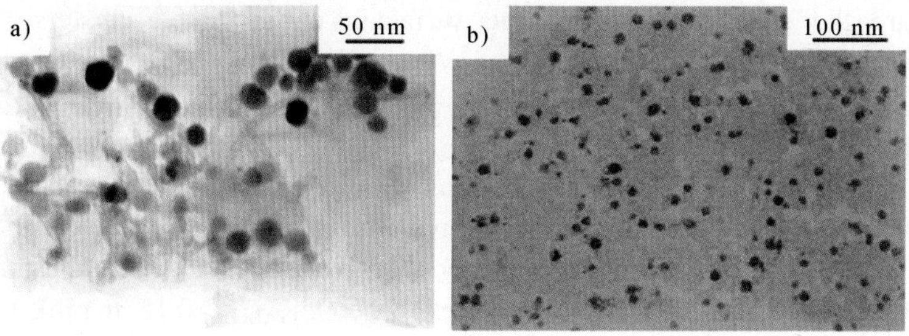

Fig.3.7. TEM images of $\{Mo_{132}\}$-$(TTA)_n$ complexes in solution at less and much higher than the stoichiometric ratio of $r_{TTA^+/\{Mo_{132}\}}$ at a) c_{TTABr} = 0.028 mg·mL^{-1}, and b) c_{TTABr} = 0.18 mg·mL^{-1}, in 0.1 mg·mL^{-1} $\{Mo_{132}\}$ solution. Reprinted from Fan, Hao, 2009a. Copyright (2009), with permission from Elsevier

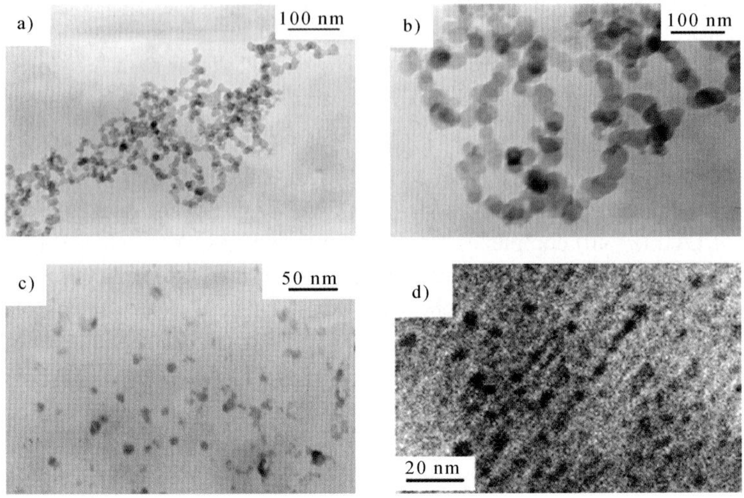

Fig.3.8. TEM images of $\{Mo_{72}Fe_{30}\}$-$(TTA)_n$ complexes at less and much higher than the stoichiometric ratio of $r_{TTA^+/\{Mo_{72}Fe_{30}\}}$ at: a), b) c_{TTABr} = 0.016 mg·mL^{-1}; c) c_{TTABr} = 0.1 mg·mL^{-1}; and d) HR-TEM image at c_{TTABr} = 0.3 mg·mL^{-1} in 0.1 mg·mL^{-1} $\{Mo_{72}Fe_{30}\}$ solution. Reprinted from Fan, Hao, 2009a. Copyright (2009), with permission from Elsevier

Based on analysis of our measurements, we schematically proposed a mechanism of the automatic and subsequent precipitation and dissolution phase transition (Fig.3.9). Taking the $\{Mo_{132}\}$-TTABr system as an example, when c_{TTABr} < 0.048 mg·mL^{-1}, *i.e.*, $n_{TTABr}:n_{\{Mo_{132}\}}$<40:1, partial negative charges of $\{Mo_{132}\}$ macroanions are neutralized by TTA$^+$ cations, so the $\{Mo_{132}\}$-$(TTA)_n$ (n<40) complexes in solution should be negatively charged. Of course, the $\{Mo_{132}\}$-$(TTA)_n$ (n<40) complexes can self-assemble to bigger complexes because of hydrophobic interactions. When c_{TTABr}= 0.048 mg·mL^{-1}, *i.e.*, $n_{TTABr}:n_{\{Mo_{132}\}}$=40:1, the negative charges of $\{Mo_{132}\}$ macroanions can be completely neutralized by TTA$^+$ cations to form hydrophobic precipitates which are the so-called hydrophobic surfactant-encapsulated clusters (HSECs). When c_{TTABr}>0.048 mg·mL^{-1}, *i.e.*, $n_{TTABr}:n_{\{Mo_{132}\}}$>40:1, hydrophobic tails of excess cationic surfactant TTA$^+$ can interact with the HSECs to form a bilayer-like structure $(TTA)_m[\{Mo_{132}\}$-$(TTA)_{40}]$ or $(TTA)_{m'}[\{Mo_{132}\}$-$(TTA)_{40}]_x$ (x>1); thus the hydrophilic heads are exposed outside of the aqueous solution. The complexes start to carry positive charges, and the amount of positive charges increases with the increase of TTABr. The precipitate complexes partly dissolve and the solution turns slightly brown. The precipitate dissolves completely when c_{TTABr} ⩾0.15 mg·mL^{-1}. Similar phase transitions of mixtures of $\{Mo_{368}\}$ and TTABr were also observed. We suggest that the automatic and subsequent precipitation and dissolution phase transition from precipitation to solution again through the interaction with different amount surfactants is reasonable.

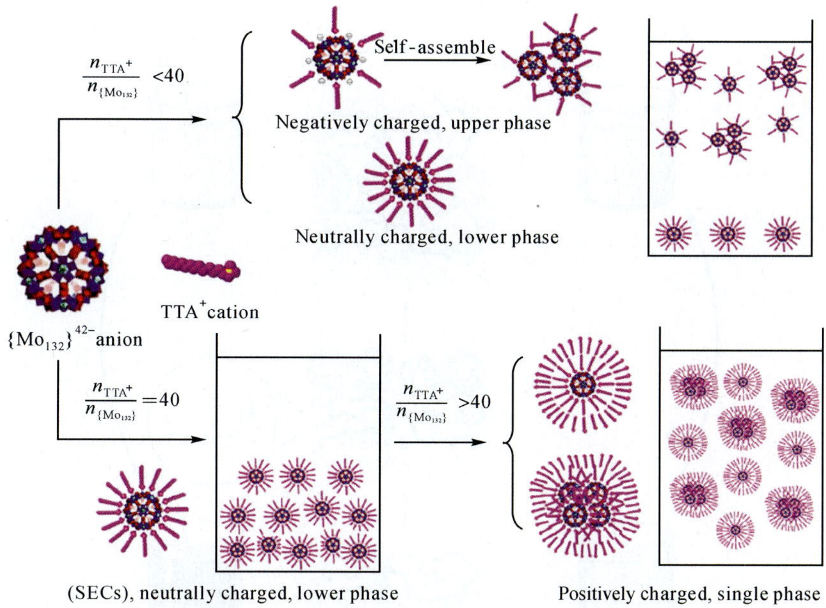

Fig.3.9. Illustration of automatic and subsequent precipitation and dissolution phase transition by the $\{Mo_{132}\}$-TTABr system as an example. Reprinted from Fan, Hao, 2009a. Copyright (2009), with permission from Elsevier

3.2.2 Multilayer Films Containing POMs by Layer-by-layer Technique on Planar Substrates

The layer-by-layer (LbL) self-assembly method based on electrostatic interactions is a simple but powerful strategy for fabricating multilayers (Decher, 1997). Developing polyoxometalates with a variety of topological and electronic properties are good candidates for constructing functional multilayers using the LbL technique. The advantage is that each nanometer-scaled POM molecule has a homogeneous diameter and surface charge when dissolved in a polar solvent.

Inorganic-organic hybrid multilayer films containing POMs assembled on planar substrates can be found elsewhere (Moriguchi, Fendler, 1998; Caruso et al., 1998a; Kurth et al., 2000b). Self-assembly of alternating layers of POM anions and oppositely charged species is deceptively simple. The LbL approach is schematically summarized in Fig.3.10, taking the electrostatic interaction of POM and polyelectrolyte (PE) as a typical example (Liu et al., 2003). For example, starting with a negatively charged substrate, such as a layer of polystyrene-sulfonate (PSS) or a charged monolayer of an alkyl-silane or alkyl-thiol, its immersion into a polycation solution leads to the adsorption of the polycations, thereby recharging the surface. Typically, a few minutes are sufficient to establish a complete layer. After rinsing the sample,

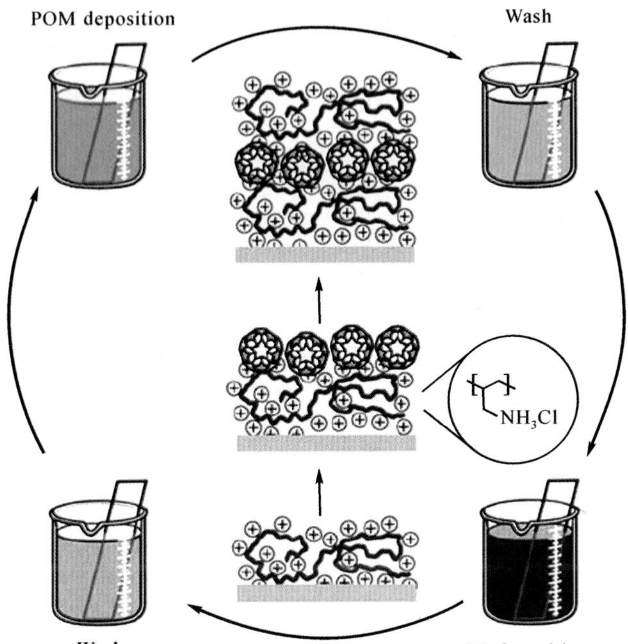

Fig.3.10. LbL self-assembly on planar substrates relies primarily on electrostatic interactions of oppositely charged adsorbents. Multilayer growth proceeds in a sequential process, in which the substrate is immersed in dilute solutions of oppositely charged species with intermittent washing steps. Combinations of different components in a single film are easily put into practice. Reprinted from Liu et al., 2003. Copyright (2003), with permission from Springer

immersion in a POM solution results in adsorption of the next layer. Repetition of this alternating deposition leads to the build-up of multilayer thin films. The only requirement is that the components are sufficiently charged in order to adsorb irreversibly at the interface.

Multilayer hybrid films composed of chitosan and Keplerate-type POM $\{Mo_{72}Fe_{30}\}$ were fabricated on quartz, silicon, and ITO substrates by the LbL method in our group (Fan, Hao, 2009b). Chitosan, poly-β-(1,4)-*D*-glucosamine (Fig.3.11) has an N-deacetylated derivative polyelectrolyte of chin and the second-most abundant natural polysaccharide after cellulose, with excellent biodegradability, biocompatibility, and nontoxicity. With a pKa of 6.2, chitosan is insoluble in alkaline and neutral solutions, but is highly positively charged because of protonation of the amino groups in acid media. Many applications of chitosan depend on the interaction between amino groups and anionic surface-active substances, such as small molecule surfactants, phospholipids, or polyoxometalate. Based on electrostatic interaction, (chitosan/$\{Mo_{72}Fe_{30}\}$)$_n$ multilayer films were fabricated and characterized on quartz slides, silicon wafers, and ITO-coated glass. These solid substrates were cleaned by immersion for 20 min at 50 °C in a series of ultrasonically agitated solvents (acetone, ethanol, H_2O). The cleaned quartz slides and silicon wafers were immersed in piranha solution (3:7, v/v 30% H_2O_2/concentrated H_2SO_4) at 80 °C for 1 h, followed by rinsing with deionized water and drying under a nitrogen stream, and then immersed in a 70 °C solution of H_2O-H_2O_2-NH_4OH (5:1:1, v/v/v) for 30 min, and washed and dried under a nitrogen stream. The ITO glass was dipped in piranha solution (3:7, v/v 30% H_2O_2/concentrated H_2SO_4) for a few minutes, and then rinsed with deionizd water and dried in a nitrogen stream. After the surface-modification, these substrates became negatively charged. Subsequently, the hydrophilic substrates were immersed in an aqueous solution of 1.0 mg·mL^{-1} chitosan (pH ~2.5, adjusted by 1.0 mol·L^{-1} hydrochloride aqueous solution) for 10 min, rinsed with washing solution, and dried under a nitrogen stream. The positively charged chitosan-coated substrates were then exposed to an aqueous solution of 1.0 mg·mL^{-1} $\{Mo_{72}Fe_{30}\}$ for 10 min to adsorb a negatively charged layer, followed by rinsing and drying. Alternate immersions in the two aqueous solutions were performed until a film with the desired number of chitosan/$\{Mo_{72}Fe_{30}\}$ multilayers was achieved. A schematic map of the LbL multilayer film of (chitosan/$\{Mo_{72}Fe_{30}\}$)$_n$ (n=2) is illustrated in Fig.3.12. The topographical characterization of (chitosan/$\{Mo_{72}Fe_{30}\}$)$_n$ (n=2) films was studied by atomic force microscopy (AFM) (Fig.3.13), which showed the distribution of aggregated nanoclusters with uniform and smooth films of $\{Mo_{72}Fe_{30}\}$ entrapped or surrounded by chitosan chains. It is also possible that $\{Mo_{72}Fe_{30}\}$ aggregated to a certain level, as the cationic polymer chitosan may have reduced the coulombic repulsion of

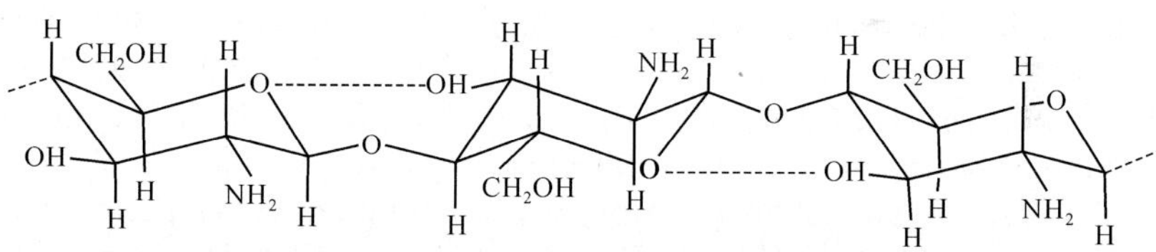

Fig.3.11. Molecular structure of chitosan. Reprinted from Fan, Hao, 2009b. Copyright (2009), with permission from ACS

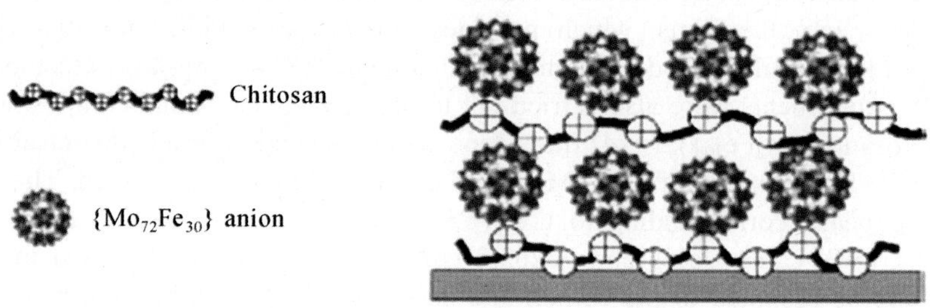

Fig.3.12. Schematic illustration of the self-assembly of a (chitosan/$\{Mo_{72}Fe_{30}\}$)$_n$ film with n=2. Reprinted from Fan, Hao, 2009b. Copyright (2009), with permission from ACS

adjacent POM centers. In addition, vertical grain structure of the multilayer surface was visible in three-dimensional AFM images, showing that the $\{Mo_{72}Fe_{30}\}$ nanoclusters embedded in the chitosan chains.

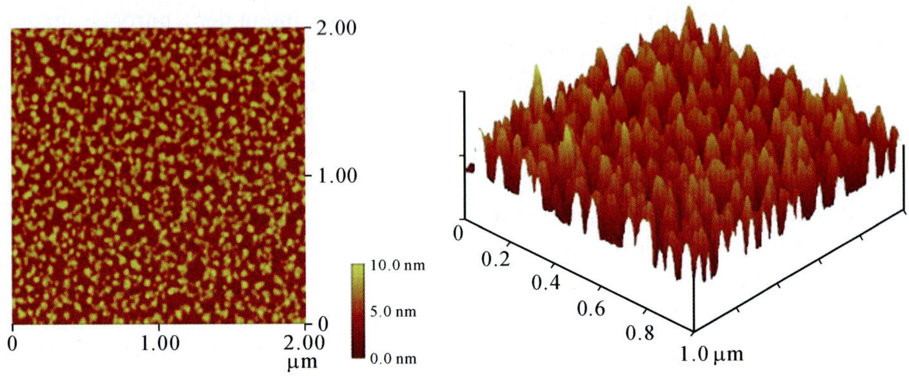

Fig.3.13. Tapping mode AFM images of (chitosan/$\{Mo_{72}Fe_{30}\}$)$_n$ (n=2) film, showing the distribution of aggregated nanoclusters with uniform and smooth films of $\{Mo_{72}Fe_{30}\}$ entrapped or surrounded by chitosan chains. Reprinted from Fan, Hao, 2009b. Copyright (2009), with permission from ACS

For LbL self-assembly on planar substrates, UV-vis measurement is usually used to monitor the deposition process. Here, we used UV-vis absorption spectra to investigate the growth of (chitosan/$\{Mo_{72}Fe_{30}\}$)$_n$ multilayer films (Fig.3.14). The solution absorption spectrum of $\{Mo_{72}Fe_{30}\}$ exhibits a broad band in the 300~400 nm range (Fig.3.14a), with a maximum at 350 nm. Fig.3.14b displays the UV-vis absorption spectra of (chitosan/$\{Mo_{72}Fe_{30}\}$)$_n$ multilayer films with n ranging from 1 to 6. Chitosan has no absorption in this area. It also shows a strong absorption band at 350 nm in the (chitosan/$\{Mo_{72}Fe_{30}\}$)$_n$ multilayer film spectra, which confirms the incorporation of $\{Mo_{72}Fe_{30}\}$ in the multilayer film. The inset presents the relationships of the absorbance at 350 nm *vs.* the layer number of multilayer films. The absorbency values increase almost linearly with the number of bilayers of the LbL films, suggesting that each adsorption cycle contributes an equal amount of $\{Mo_{72}Fe_{30}\}$ into the films, which provides persuasive evidence for the regular growth of the multilayer and for high reproduction of the layer-by-layer assembly.

XPS experiments were also carried out to identify the elemental composition of the (chitosan/$\{Mo_{72}Fe_{30}\}$)$_n$ (n=4) films deposited on a single-crystal silicon substrate (Fig.3.15). The presence of C, N, Mo, and Fe in the films was confirmed. The films exhibited peaks corresponding to C1s (BE = 284.8 eV), N1s (BE = 398.4 eV), Mo3d5/2 (BE = 232.5 eV), and Fe2p3/2 (BE = 711.9 eV). The C1s and the N1s signal can be assigned to the carbon and amido in chitosan, while the Mo3d and Fe2p are ascribed to the Keplerate-type $\{Mo_{72}Fe_{30}\}$ molecule. XPS results again confirm the existence of cationic chitosan and $\{Mo_{72}Fe_{30}\}$ polyanions in the multilayer films in conjunction with the results of UV-vis spectra.

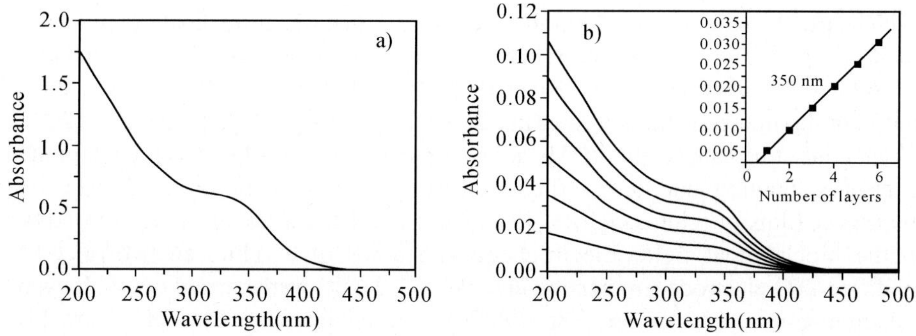

Fig. 3.14. UV-vis absorption spectra of a) 0.05 mg·mL^{-1} $\{Mo_{72}Fe_{30}\}$ aqueous solution, and b) (chitosan/$\{Mo_{72}Fe_{30}\}$)$_n$ multilayer films with n = 1~6 on quartz substrates. The curves, from bottom to top, are correspond to n = 1, 2, 3, 4, 5, and 6. The inset shows the relationships of the absorbance at 350 nm *vs.* the layer number of multilayer films. Reprinted from Fan, Hao, 2009b. Copyright (2009), with permission from ACS

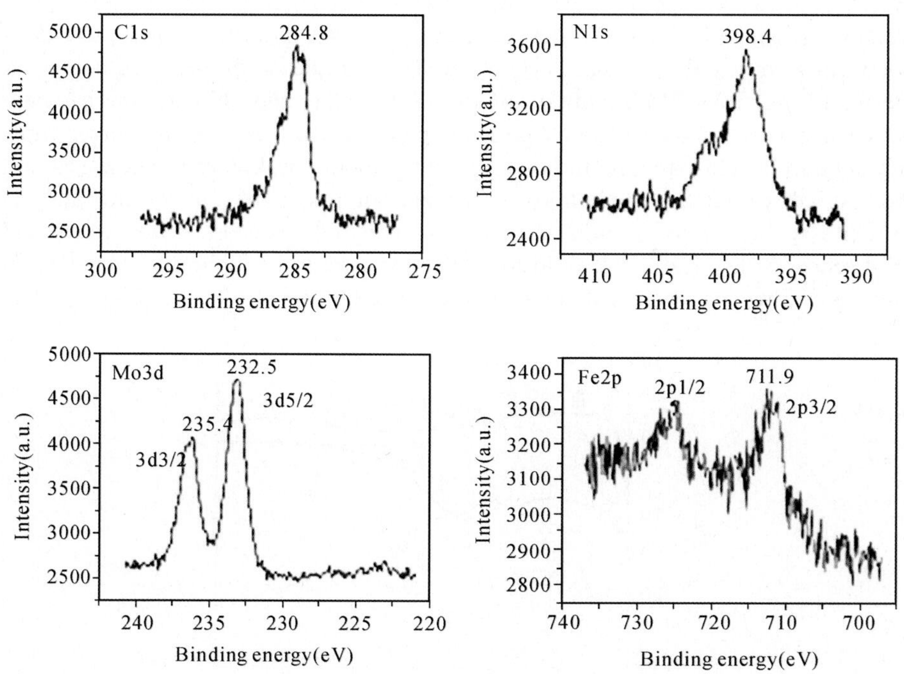

Fig. 3.15. XPS spectra of the multilayer film (chitosan/$\{Mo_{72}Fe_{30}\}$)$_n$ (n=4). Reprinted from Fan, Hao, 2009b. Copyright (2009), with permission from ACS

The combining of functional POMs having a variety of topological and electronic properties makes the self-assembled films valuable. Kurth's group reported a device design based on an electrostatic complex of a POM cluster and a polyelectrolyte (Liu et al., 2006b). They found that the film coating reversibly changed color from transparent to blue by photo- and/or electro-induced stimulation.

Some other electrochromic (Gao et al., 2005a), photochromic (Chen et al., 2004), and photoluminescent films (Xu et al., 2002; Chen et al., 2001; Jiang et al., 2005), which consist of nanoscopic POMs are also attractive. $\{Mo_{72}Fe_{30}\}$ can be exploited in electrocatalytic reductions, and chitosan is a typical biologic molecule. Thus the multilayer films of (chitosan/$\{Mo_{72}Fe_{30}\}$)$_n$ are predicted to serve as electrocatalysis in the biochemical field. Electrochemical studies of a (chitosan/$\{Mo_{72}Fe_{30}\}$)$_n$-modified ITO electrode (immersion area 0.6 × 1.0 cm^2) were carried out with cyclic voltammograms (CVs) (Fan, Hao, 2009b). The comparative CVs of a bare ITO electrode just after immersion into pH 1.5 Na_2SO_4-H_2SO_4 solution containing 0.1 mg·mL^{-1} $\{Mo_{72}Fe_{30}\}$, and a multilayer film (chitosan/$\{Mo_{72}Fe_{30}\}$)$_4$-modified ITO electrode in the same buffer solution, are shown in Fig.3.16. In the range −0.4 to 0.8 V, the thin film underwent three redox waves with main peak potentials of 0.15, −0.17, and −0.22 V for peaks I, II and III, corresponding to three two-electron reversible transfer processes. Moreover, the CVs for the electrocatalytic reduction of iodate, chlorate, and bromate by (chitosan/$\{Mo_{72}Fe_{30}\}$)$_4$ in pH 1.5 Na_2SO_4-H_2SO_4 solution are given in Fig.3.17. The reduction of iodate, bromate, and chlorate does not occur significantly on a bare ITO electrode under the conditions used below for the multilayer films. With addition of iodate, bromate, and chlorate, the cathodic peak current of the second redox peak of (chitosan/$\{Mo_{72}Fe_{30}\}$)$_4$ multilayer films substantially increased while the corresponding anodic peak current decreased and the first redox peak remained almost unchanged. In other words, the electrocatalysis of (chitosan/$\{Mo_{72}Fe_{30}\}$)$_4$ multilayer films towards iodate occurs at the second and third redox peaks. The increase in the reduction current of (chitosan/$\{Mo_{72}Fe_{30}\}$)$_4$ multilayer films was most distinct when the concentration of IO_3^- was higher. Also,

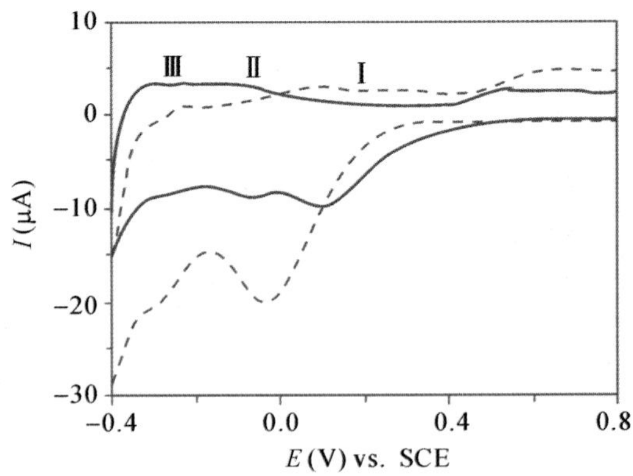

Fig.3.16. Comparative cyclic voltammograms of 0.1 mg·mL^{-1} $\{Mo_{72}Fe_{30}\}$ (dotted lines, ITO electrode), and a (chitosan/$\{Mo_{72}Fe_{30}\}$)$_4$ multilayer film-modified ITO electrode (solid lines) in pH 1.5 Na_2SO_4-H_2SO_4 solution. Scan rate: 100 mV·s^{-1}. Reprinted from Fan, Hao, 2009b. Copyright (2009), with permission from ACS

the multilayer films exhibited electrocatalytic reduction activity of $IO_3^- > BrO_3^- > ClO_3^-$ (Fig.3.17). Therefore, $\{Mo_{72}Fe_{30}\}$ acts as a very efficient electron transfer system in the electrocatalytic reduction process. The electrocatalytic activity of (chitosan/$\{Mo_{72}Fe_{30}\}$)$_4$ multilayer films are reproducible due to the good stability of the hybrid films. These results show that chitosan can be used effectively as a film-forming material or a medium to carry functional substances such as POMs. And the films are expected to have potential applications in the field of functional materials, electrocatalysis, and POM-based medicine science.

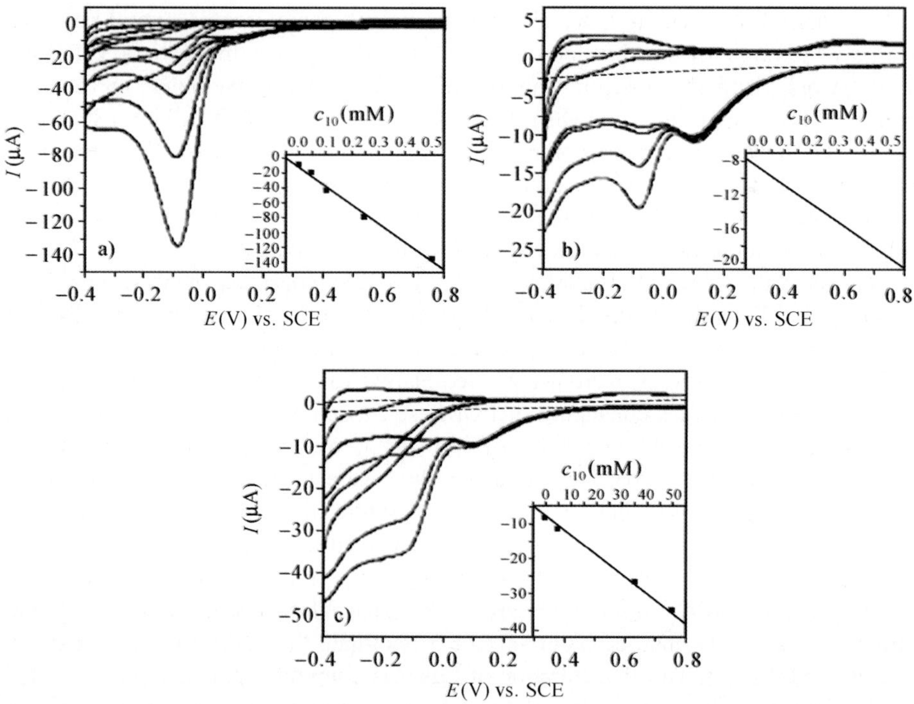

Fig.3.17. Cyclic voltammograms of ITO/(chitosan/$\{Mo_{72}Fe_{30}\}$)$_4$ multilayer films in pH 1.5 Na_2SO_4-H_2SO_4 buffer solutions with various concentrations of anions. a) [IO_3^-]: 0.0, 0.05, 0.075, 0.1, 0.25, and 0.5 mM (solid lines). Also, bare ITO electrode and [IO_3^-] = 0.5 mM (dotted lines). b) [BrO_3^-]: 0.0, 0.05, 0.25, and 0.5 mM (solid lines). Also, bare ITO electrode and [BrO_3^-] = 0.5 mM (dotted lines). c) [ClO_3^-]: 0.0, 5, 35, and 50 mM (solid lines). Also, bare ITO electrode and [ClO_3^-] = 50 mM (dotted lines). Scan rate: 100 mV·s^{-1}. Insets show the relationship between catalytic currents and concentrations of IO_3^-, BrO_3^-, and ClO_3^- anions. Reprinted from Fan, Hao, 2009b. Copyright (2009), with permission from ACS

3.2.3 Multilayer Films Containing POMs by Layer-by-layer Technique into Spherical Nanocapsules

Besides LbL self-assembly on planar substrates, in the past 10 years the LbL self-assembly method has also been widely applied to fabricate micro-polyelectrolyte (PE) capsules using micro-spheres as cores, and this is also based on electrostatic interactions (Donath et al., 1998; Sukhorukov et al., 2004). Various templates have been used as cores and different components for assembling into the capsule shells. Recently, substantial progress has been made in extending the capsules to functional ones through fabricating inorganic components into PE capsules (Shchukin et al., 2005; Gaponik et al., 2004), which provide the inorganic-organic hybrid capsules with a wide range of novel properties, such as luminescence, magnetism, catalysis, and sensors. Typical inorganic nanoparticles that can be assembled in capsules include SiO_2 (Caruso et al., 1998b), Fe_3O_4 (Shchukin et al., 2004), TiO_2 (Shchukin et al., 2003a), and rare earth compounds (Shchukin et al., 2003b). Very recently, combining the POM chemistry and capsule fields, Wang's group has studied the self-assembly of POM into microcapsules (Gao et al., 2005b). They used $Na_9EuW_{10}O_{36}$ molecules to endow capsules with fluorescence.

We chose the fully hydrophilic $\{Mo_{72}Fe_{30}\}$ as the homogeneous nano-size particles, PEs of poly(allylamine hydrochloride) (PAH), and poly(sodium 4-styrenesulfonate) (PSS) as shell elements to fabricate capsules (Cui et al., 2009). Poly (styrene-acrylic acid) (PSA) latex particles were used as templates. Taking this as a typical example to discuss nanocapsules consisting of POMs self-assembled by the LbL technique, the magnetic nanocapsules consisting of $\{Mo_{72}Fe_{30}\}$ were fabricated as follows. The negatively charged PSA particles as a template were incubated with 1 mL of the positively charged PAH solution at room temperature for 20 min, followed by three cycles of centrifugation/rinsing, and finally dispersed in water. The PSS layer was assembled in the same way. The PAH and PSS adsorption steps were repeated to build $(PAH/PSS)_2/PAH$ multilayers on the PSA particles to produce a smooth and uniformly positively charged outer surface to facilitate the adsorption of negatively charged $\{Mo_{72}Fe_{30}\}$. The concentration of PEs was 2 $mg \cdot mL^{-1}$ containing 0.5 $mol \cdot L^{-1}$ NaCl. Subsequently, $\{Mo_{72}Fe_{30}\}$/PAH alternate adsorption was performed on $(PAH/PSS)_2/PAH$ nanocapsules using 2 $mg \cdot mL^{-1}$ $\{Mo_{72}Fe_{30}\}$ containing 0.1 $mol \cdot L^{-1}$ NaCl with a pH 3.5. After adsorption of each layer, the particles were washed 3 times with distilled water and centrifuged at 2,000 *g* for 5 min. For $\{Mo_{72}Fe_{30}\}$-embedded hollow nanocapsules, the template PSA core was dissolved in tetrahydrofuran. Preparation of the $\{Mo_{72}Fe_{30}\}$-embedded nanocapsules self-assembled on PSA is schematically illustrated in Fig.3.18.

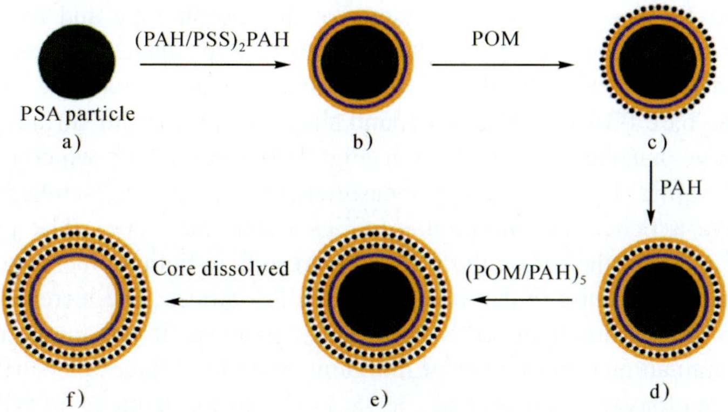

Fig.3.18. LbL assembly of PAH, PSS, and POM ({$Mo_{72}Fe_{30}$}) on PSA particles and hollow microcapsules. a)~b) The first stage involves stepwise deposition of PEs on PSA colloidal particles. b)~e) Alternate assembly PAH and {$Mo_{72}Fe_{30}$} nanoparticles. e)~f) Decomposition of PSA cores and formation of hollow [(PAH/PSS)$_2$(PAH/{$Mo_{72}Fe_{30}$})$_6$PAH] microcapsules. Reprinted from Cui et al., 2009. Copyright (2009), with permission from Elsevier

PSA template particles are milky white latexes dispersed in water (Fig.3.19a). They have a narrow size distribution with an average diameter of 450 nm from dynamic light scattering and SEM measurements. The 2.5 nm diameter, C_{60}-like {$Mo_{72}Fe_{30}$} molecules behave like a weak acid and are a transparent pale yellow (Fig.3.19b) in aqueous solution due to the partial deprotonation of the water ligands chemically attached to their Fe^{III} centers, thus making the {$Mo_{72}Fe_{30}$} clusters slightly negatively charged (carrying several localized charges), which agrees very well with the requirement of LbL self-assembly. A photograph of {$Mo_{72}Fe_{30}$}-embedded nanocapsules dispersed in water is shown in Fig.3.19c. Macroscopic observation can be the most intuitive and simple method of investigating the changes during experiments. The color changes of latex from milky white to pale yellow confirm that {$Mo_{72}Fe_{30}$} has been self-assembled in the nanocapsule shells.

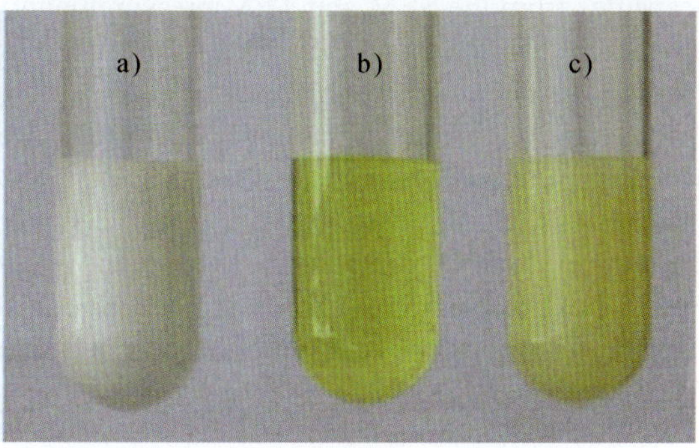

Fig.3.19. Photographs of a) PSA latex, b) {$Mo_{72}Fe_{30}$} solution with a concentration of 2 mg·mL^{-1}, and c) [(PAH/PSS)$_2$(PAH/{$Mo_{72}Fe_{30}$})$_6$PAH] nanocapsules dispersed in water. Reprinted from Cui et al., 2009. Copyright (2009), with permission from Elsevier

SEM and TEM were used to characterize the morphology and the multilayer growth process of hybrid nanocapsules. Representative images of bare PSA latex particles and $[(PAH/PSS)_2(PAH/\{Mo_{72}Fe_{30}\})_6PAH]$ nanocapsules are displayed in Fig.3.20. The bare PSA cores have a round shape, and a smooth surface and edges with an average diameter of (450±10) nm (Figs.3.20 a and 3.20d), which is consistent with the dynamic light scattering measurements. $\{Mo_{72}Fe_{30}\}$-embedded nanocapsules have a lichee-like morphology (Figs.3.20b and 3.20e). The presence of PEs-$\{Mo_{72}Fe_{30}\}$ multilayers on the PSA cores results in both an increase in surface roughness and an increase in the diameter of the PS lattices. The increase in surface roughness is mainly due to adsorbed $\{Mo_{72}Fe_{30}\}$ particles. It was found that adsorption of PEs onto an outermost layer of inorganic particles reduces the surface roughness of the multilayer (Caruso et al., 1998c). The mean diameter of $[(PAH/PSS)_2(PAH/\{Mo_{72}Fe_{30}\})_6PAH]$ nanocapsules is about 530 nm, which yields an increase in diameter of 80 nm over that of the PSA latex particles. According to single- particle light scattering and AFM measurements, the monolayer thickness of PE (either PAH or PSS) is about 1.5 nm (Sukhorukov et al., 1998; Leporatti et al., 2000). So the thickness of every PAH/$\{Mo_{72}Fe_{30}\}$ bilayer is about 12 nm, which is smaller than the monolayer thickness of inorganic hollow microcapsules (Caruso, 2000). The reason may be that the diameter of $\{Mo_{72}Fe_{30}\}$ is small and it can be assembled in the shell more tightly. In this case, shell thickness can be tailored at the nanometer level more precisely.

After dissolving the core in tetrahydrofuran, the nanocapsules became hollow spheres with hybrid shells composed of $\{Mo_{72}Fe_{30}\}$ and polyelectrolytes (Fig.3.20). The PAH bridged the $\{Mo_{72}Fe_{30}\}$ nanoparticles in the shell tightly. The shell thickness of hollow nanocapsules determined from the TEM images was (80±5) nm, which is consistent with the calculation result of nanocapsules assembled on PSA particles. The $\{Mo_{72}Fe_{30}\}$-embedded hollow nanocapsules have an apple-like shape in vacuum (Figs.3.20c and 3.20f). The reason may be that the hybrid shell is not invariably sufficient to maintain the initial spherical structure of the PSA latex particle and collapses at thin places. However, no holes of rupture were identified in the hollow nanocapsules from the SEM and TEM measurements, which indicate that the hollow spheres were intact. So the $\{Mo_{72}Fe_{30}\}$-embedded hollow nanocapsules can be potentially used as containers or nanoreactors.

In order to characterize the elemental composition of the nanocapsules, EDS was used for the $\{Mo_{72}Fe_{30}\}$-embedded nanocapsules. EDS analysis revealed that the atomic ratio of Mo:Fe was 43:18, nearly equal to 78:30 according to the molecular formula of $\{Mo_{72}Fe_{30}\}$, demonstrating that $\{Mo_{72}Fe_{30}\}$ existed in the hollow nanocapsules (Fig.3.21). The element O was mainly introduced from $\{Mo_{72}Fe_{30}\}$. The elements N and S were introduced from the PAH and PSS, respectively.

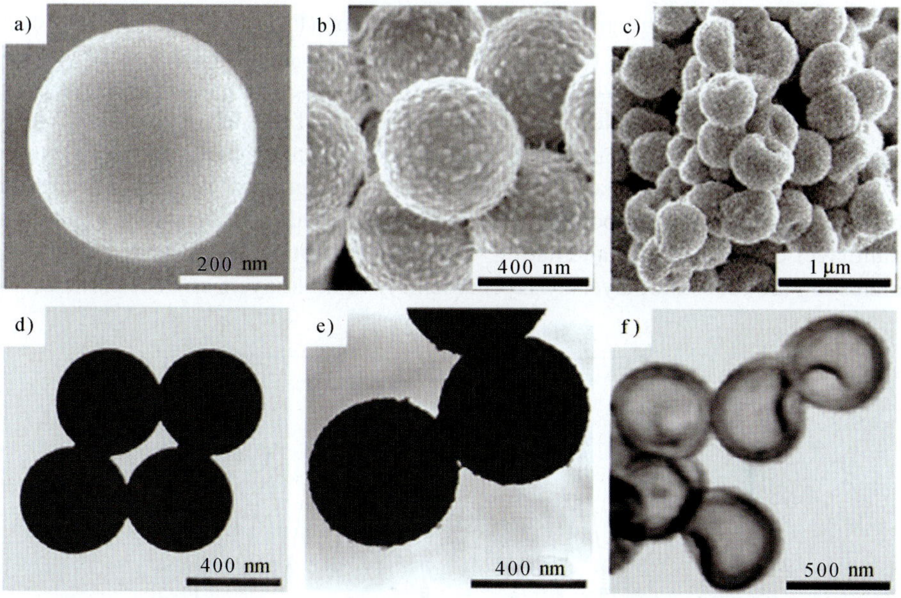

Fig.3.20. a)~c) SEM images; d)~f) TEM images. a), d) PSA latex particles; b), e) $[(PAH/PSS)_2(PAH/\{Mo_{72}Fe_{30}\})_6PAH]$ nanocapsules self-assembled on PSA; c), f) hollow $[(PAH/PSS)_2(PAH/\{Mo_{72}Fe_{30}\})_6PAH]$ nanocapsules. Reprinted from Cui, et al. 2009. Copyright (2009), with permission from Elsevier

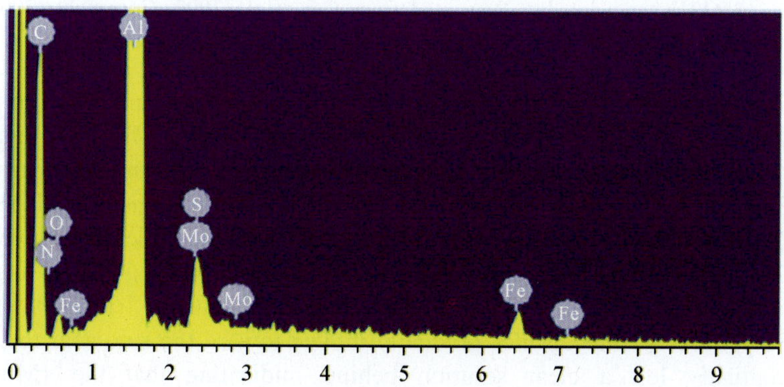

Fig.3.21. EDS analysis of hollow $[(PAH/PSS)_2(PAH/\{Mo_{72}Fe_{30}\})_6PAH]$ nanocapsules. Reprinted from Cui et al., 2009. Copyright (2009), with permission from Elsevier

In addition, we know that the $[(PAH/PSS)_2(PAH/\{Mo_{72}Fe_{30}\})_6PAH]$ nanocapsules were prepared by the alternate adsorption of positive PAH and negative $\{Mo_{72}Fe_{30}\}$ on a precursor multilayer $(PAH/PSS)_2$ assembled onto negatively charged PSA latex particles. So the ζ-potential of the nanocapsules during the self-assembly was monitored to indicate the growth of multilayer films (Fig.3.22). The odd and even layer numbers correspond to the PAH and $\{Mo_{72}Fe_{30}\}$ adsorption steps, except for layers number 2 and 4 which correspond to PSS adsorption. The layer number 0 corresponds to PSA particles. The surface potential of the nanocapsules is highly negatively charged. This could be because of incomplete coverage by the first layer PAH with an excess of negative charge and plentiful carboxyl on the surface of the PSA latex particles. The ζ-potential measurements revealed the surface charge of the multilayer coated particles alternating with each adsorption of PEs and $\{Mo_{72}Fe_{30}\}$, indicating the successful layer-by-layer assembly of multilayers on PSA particles.

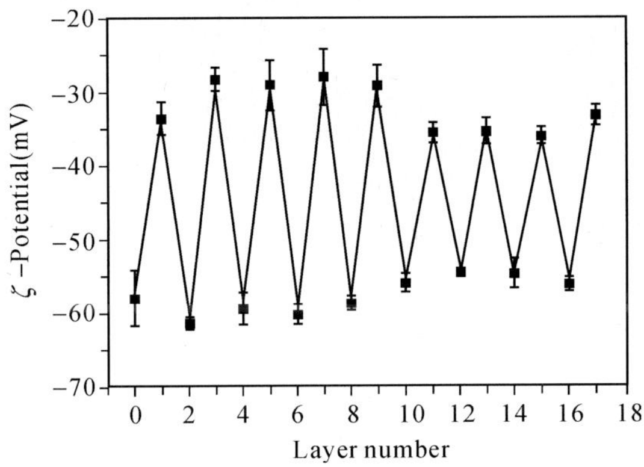

Fig.3.22. ζ-potential as a function of the number of layers in $[(PAH/PSS)_2(PAH/\{Mo_{72}Fe_{30}\})_6PAH]$ nanocapsules. Reprinted from Cui et al., 2009. Copyright (2009), with permission from Elsevier

Müller and coworkers reported that $\{Mo_{72}Fe_{30}\}$ is the largest paramagnetic molecule. Interestingly and surprisingly, we found that $\{Mo_{72}Fe_{30}\}$-embedded nanocapsules could be separated in water by a magnet but the nanoparticles of $\{Mo_{72}Fe_{30}\}$ in the solution could not be separated. The response of $[(PAH/PSS)_2(PAH/\{Mo_{72}Fe_{30}\})_6PAH]$ nanocapsules to an external magnetic field was investigated (Fig.3.23). The nanocapsules dispersed in water were in homogeneous suspension before applying a magnetic field. However, the diluted $[(PAH/PSS)_2(PAH/\{Mo_{72}Fe_{30}\})_6PAH]$ nanocapsules gradually moved toward the side near the magnet and eventually left a clear solution behind, indicating that the $\{Mo_{72}Fe_{30}\}$-embedded nanocapsules were magnetically active. A possible explanation is that the $\{Mo_{72}Fe_{30}\}$ molecules embedded in the nanocapsules were magnetized by the external field, causing the nanocapsules to become magnetically active. However, the solution of $\{Mo_{72}Fe_{30}\}$ is a thermodynamically stable system. The external magnetic field could not destroy the balance of the system. Moreover, the behavior of the $\{Mo_{72}Fe_{30}\}$-embedded nanocapsules under an external magnetic field was examined by TEM. When a drop of diluted $[(PAH/PSS)_2(PAH/\{Mo_{72}Fe_{30}\})_6PAH]$ nanocapsules was applied and allowed to evaporate in the presence of a magnetic field, nanocapsules aligned in parallel stripes were obtained (Fig.3.24). This also revealed that the $\{Mo_{72}Fe_{30}\}$-embedded nanocapsules show a good response to magnetic fields. The nanocapsules can be manipulated by external magnetic fields.

Fig.3.23. Photograph showing the separation of $[(PAH/PSS)_2(PAH/\{Mo_{72}Fe_{30}\})_6PAH]$ nanocapsules after applying a magnet to the suspension. Reprinted from Cui et al., 2009. Copyright (2009), with permission from Elsevier

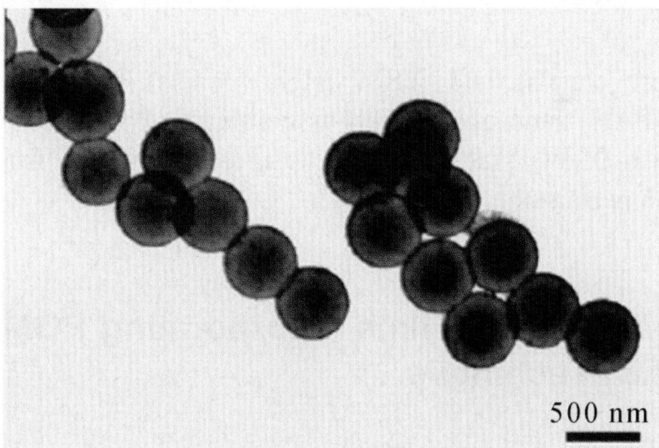

Fig.3.24. TEM image of $\{Mo_{72}Fe_{30}\}$-embedded nanocapsules aligned in an external magnetic field. Reprinted from Cui et al., 2009. Copyright (2009), with permission from Elsevier

As well as C_{60}-like "Keplerate"-type $\{Mo_{72}Fe_{30}\}$, sandwich-type $\{Mn_2Bi_2W_{20}\}$ with the molecular formula $Na_6(NH_4)_4[(Mn^{II}(H_2O)_3)_2(WO_2)_2(BiW_9O_{33})_2]\cdot 37H_2O$ (Bösing et al., 1998) was used to fabricate nanocapsules by the LbL technique, and similar nanocapsules were obtained (Fig.3.25).

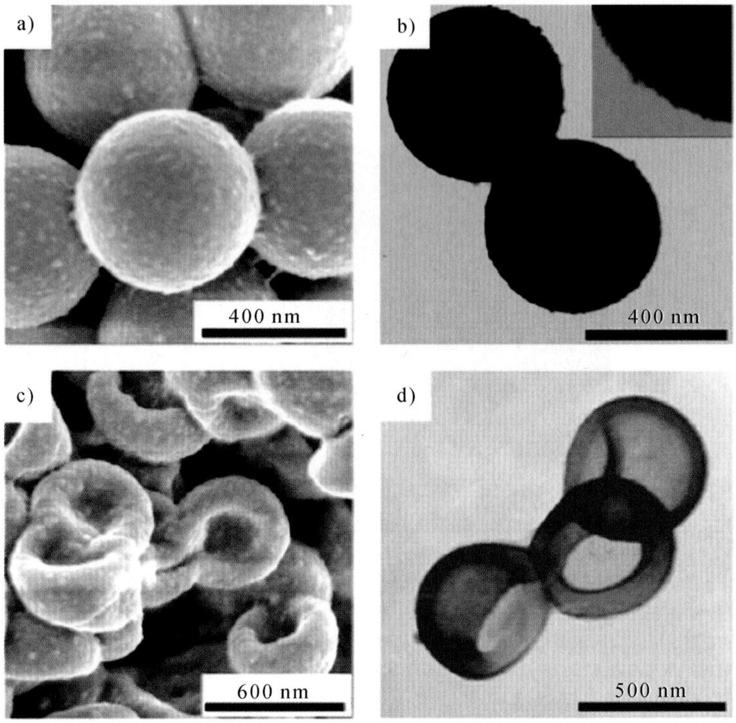

Fig.3.25. SEM (left panels) and TEM (right panels) images: a), b) $[(PAH/PSS)_2(PAH/\{Mn_2Bi_2W_{20}\})_6PAH]$ nanocapsules self-assembled on PSA latex particles; c), d) hollow $[(PAH/PSS)_2(PAH/\{Mn_2Bi_2W_{20}\})_6PAH]$ nanocapsules. Reprinted from Cui et al., 2009. Copyright (2009), with permission from Elsevier

3.2.4 Monolayer/Multilayer Films Incorporating POMs by Langmuir-Blodgett (LB) Technique

The old but elegant Langmuir-Blodgett (LB) technique is clearly one way to arrange molecules into organized assemblies. Generally, some organic molecules are used and easy to fabricate into ordered multilayer films by the LB method. To meet more requirements, it is necessary to construct the functional films by incorporating inorganic species. Recently, organic-inorganic hybrid monolayer and/or multilayer films composed of POMs and surfactants have been fabricated by the LB technique. By taking advantage of the adsorption properties of giant anions of POMs along a positively charged monolayer spread in water, Mingotaud et al. successfully obtained the first organic-inorganic LB films following the method schematized in Fig.3.26 (Coronado, Mingotaud, 1999). This consists of repetitive dipping of a solid substrate through a compressed monolayer spread at the gas/water interface. During the up and down strokes, repetitive transfer of compressed monolayers onto a solid substrate leads to homogeneous multilayers, producing a material with a precise thickness and a lamellar structure. The schematic structure of an LB multilayer film

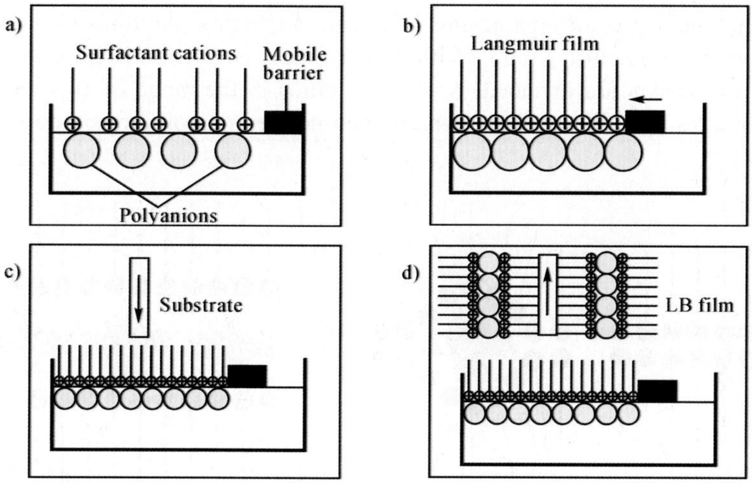

Fig.3.26. Steps for building up an LB film containing monolayers of POM clusters. Reprinted from Coronado, Mingotaud, 1999. Copyright (1999), with permission from Wiley

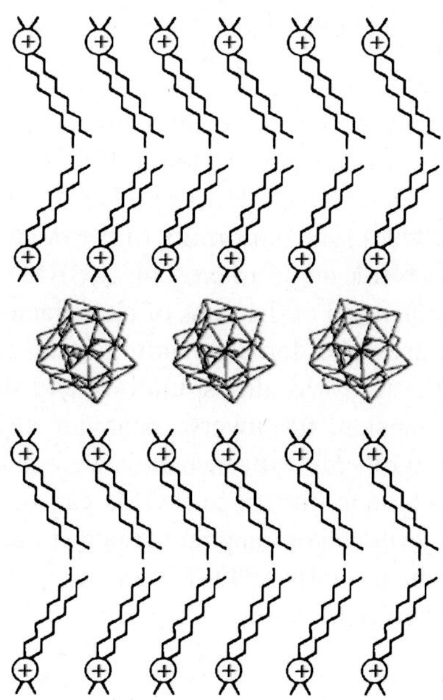

Fig.3.27. Schematic structure of the LB film composed of POMs and double-chain surfactants. Reprinted from Coronado, Mingotaud, 1999. Copyright (1999), with permission from Wiley

containing POMs and double-chain surfactants is shown in Fig.3.27. Using the LB technique, a magnetic multilayer film based on the single-molecule magnet Mn_{12} has also been reported (Clemente-León et al., 1998b). Furthermore, these workers concluded that the degree of organization of the clusters within the films is strongly dependent on the lipid:cluster ratio. With higher ratios, isolated clusters or partial monolayers of clusters are obtained, while with ratios in the range of 10:1 to 5:1, lamellar structures with clusters organized in well-defined monolayers are observed (Fig.3.28).

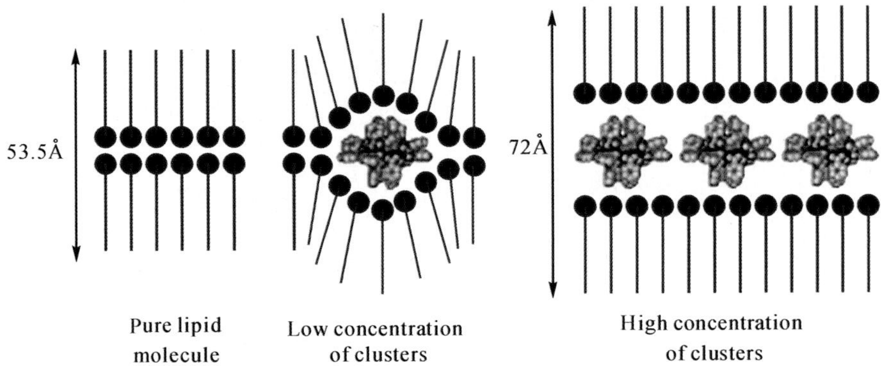

Fig.3.28. Schematic structures of the LB films formed from Mn_{12} clusters and lipids when the lipid:cluster ratio is varied. Reprinted from Clemente-León et al., 1998b. Copyright (1998), with permission from Wiley

The Langmuir monolayer film of the supramolecular architecture of nanoporous hydrophobic surfactant-encapsulated clusters (HSECs) with the empirical formula $(DODA)_{40}(NH_4)_2[(H_2O)_n{\cdot}Mo_{132}O_{372}(CH_3CO_2)_{30}(H_2O)_{72}]$ ($n \approx 50$) has been reported in detail (Kurth et al., 2000c). The uniformity of the monolayer at different pressures was investigated by Brewster angle microscopy (BAM) (Fig.3.29). The gas phase region at low pressure consisted of domains of uncharacteristic shape and size. Upon decreasing the surface area, the domains started to fuse (0 $mN{\cdot}m^{-1}$). At a surface pressure of 5 $mN{\cdot}m^{-1}$ the film was almost uniform and stayed like this upon further compression. Upon expansion, the inverse behavior occurred: the Langmuir film remained uniform up to 0 $mN{\cdot}m^{-1}$, after which it segregated into domains.

By using the LB technique, functional POMs can be effectively assembled into ordered Langmuir monolayer and/or Langmuir-Blodgett multilayer films. Consequently, in future, these films incorporating POMs can be applied in catalysis, medicine, coating, and electronic devices.

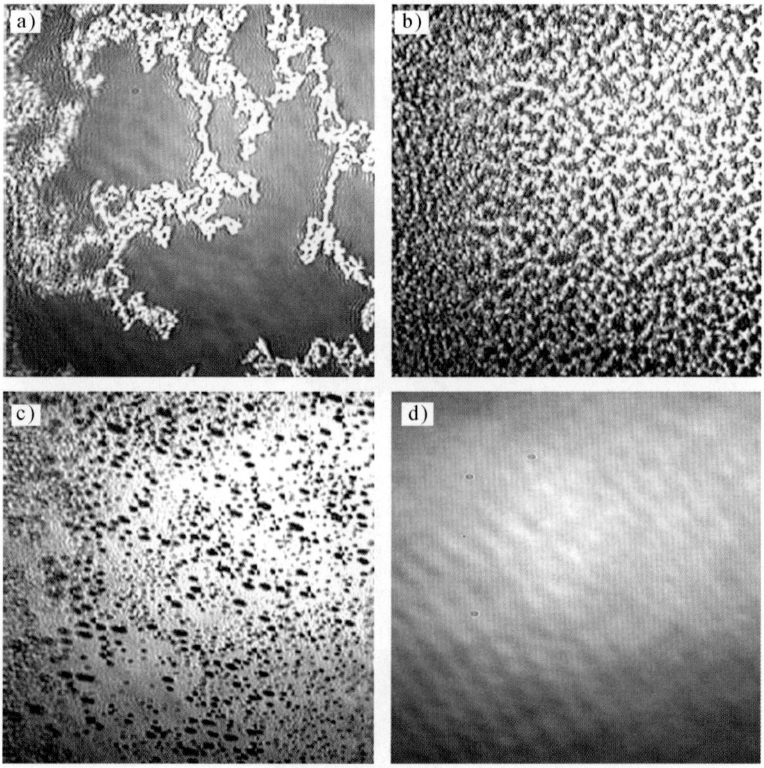

Fig.3.29. BAM images of the Langmuir monolayer of $(DODA)_{40}(NH_4)_2[(H_2O)_n \cdot Mo_{132}O_{372}(CH_3CO_2)_{30}(H_2O)_{72}]$ ($n \approx 50$) (15 °C, pure water, image area 500 μm × 500 μm). Images a)~c) were recorded at zero surface pressure at a) 34, b) 31, and c) 29 nm^2 per HSEC. Image d) was recorded at a surface pressure of 5 $mN \cdot m^{-1}$. After the spreading of this HSEC cluster, film domains of uncharacteristic shape and size appeared, which began to fuse upon compression. At a surface pressure of 5 $mN \cdot m^{-1}$ the monolayer became homogeneous (compression direction from bottom to top of the image). Reprinted from Kurth et al., 2000c. Copyright (2000), with permission from the RSC Publishing

3.2.5 Three-dimensional Aggregates of POM-surfactant Hybrids

As described above, functional nanoscopic polyoxometalates can be self-assembled into two-dimensional thin films by the LbL and/or LB techniques. Recently, ordered three-dimensional aggregates composed of POMs and surfactants have also been studied. Antonietti et al. found a liquid crystal of POM-surfactant hybrids (Polarz et al., 2001). A wheel-shaped $\{Mo_{176}\}$ and cationic double-chain surfactant, dioctadecyldimethylammonium (DODMA), were used. Their results are summarized in the structural scheme shown in Fig.3.30. POMs as model systems for charged inorganic colloids can be complexed or counterion-exchanged with surfactants. DODMA is strongly bound to the equator region of the ring-shaped $\{Mo_{176}\}$. Subsequently, through supramolecular interactions, the POM-surfactant units self-assembled to columns which arranged themselves into a hexagonal array.

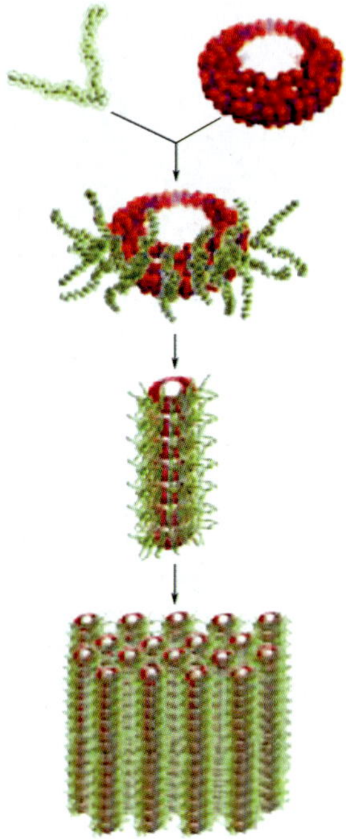

Fig.3.30. Schematic representation of the liquid crystal analogous behavior of POM-surfactant hybrids. The surfactant (shown is the DODMA cation) is strongly attached to the ring-shaped $\{Mo_{176}\}$, mainly by electrostatic forces. For the sake of clarity, only ten surfactant molecules bound to the front of the POM are shown, instead of the actual 32 located all over the periphery of the ring. Through supramolecular interactions, the POM-surfactant units self-assemble into columns which themselves arrange into a hexagonal array. Reprinted from Polarz et al., 2001. Copyright (2001), with permission from Wiley

To the best of our knowledge, surfactant-encapsulated POM clusters made by the encapsulation of cationic surfactants, such as dioctadecyldimethylammonium chloride (DODMACl), didodecyldimethylammonium chloride (DDDMACl), cetyltrimethylammonium bromide (CTABr), and tetradecyltrimethylammonium bromide (TTABr), cannot display the liquid crystalline array by thermotropic modulation. Very recently, Wu et al. have delicately fabricated stable and reversible thermotropic liquid crystals of POM-surfactant hybrids by using a series of biphenyl-containing ammonium amphiphiles with two alkyl chains, *N,N*-di[10-[4-(4′-alkyloxybiphenyl)oxy] decyl]-*N,N*-dimethylammonium bromide ($C_nBphC_{10}N$, $n = 6, 8, 10, 12$), and *N*-[12-(4-carboxylphenoxy)dodecyl]-*N*-dodecyl-*N,N*-dimethylammonium bromide (CDDA) to cap the POMs (Li et al., 2008; Yin et al., 2009). Their results of differential scanning calorimetry (DSC), polarized optical microscopy (POM), and X-ray diffraction (XRD) revealed that these complexes underwent smectic mesophases during the heating and cooling cycles. Polarized optical microscopy images (Fig.3.31) of the surfactant CDDA and the surfactant-encapsulated polyoxometalate clusters: $(CDDA)_8H(EuW_{10}O_{36})(H_2O)$, $(CDDA)_9H_2[Eu(PW_{11}O_{39})_2](H_2O)_5$, and $(CDDA)_{11}H_2[Eu(SiW_{11}O_{39})_2](H_2O)_5$ were investigated during the cooling process, from which their textures were clearly seen and the different smectic phases were indicated. It also interestingly represented a new type of liquid crystal material.

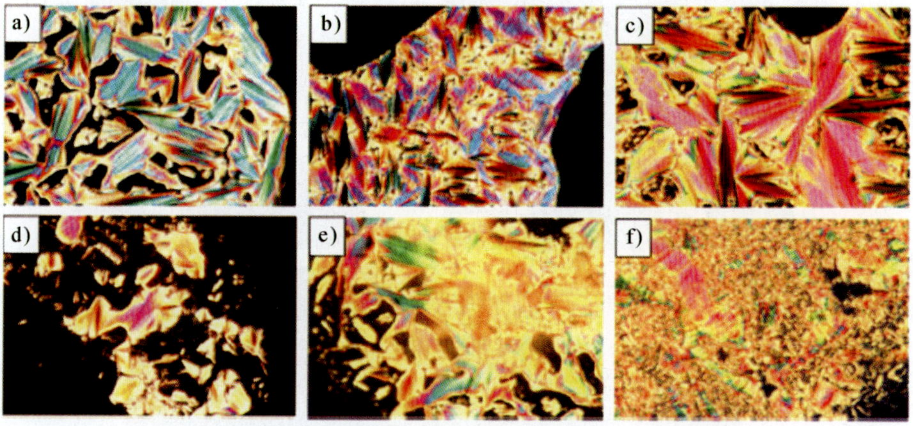

Fig.3.31. Polarized optical microscopy images of the surfactant CDDA at a) 160 °C and b) 130 °C; and the surfactant-encapsulated POM clusters: $(CDDA)_8H(EuW_{10}O_{36})(H_2O)_3$ at c) 148 °C, $(CDDA)_9H_2[Eu(PW_{11}O_{39})_2](H_2O)_5$ at d) 140 °C, and $(CDDA)_{11}H_2[Eu(SiW_{11}O_{39})_2](H_2O)_5$ at e) 138 °C and f) 120 °C during the cooling process (magnification: ×400). Reprinted from Yin et al., 2009. Copyright (2009), with permission from ACS

After the surface modification of POMs by surfactants, Wu et al. also found that the formed hydrophobic surfactant-encapsulated clusters (HSECs) have the general tendency to assemble into spherical aggregates in organic solvents, which is attributed to the rearrangement of surfactants on the exterior of the POM (Li et al., 2007). The self-assembly process can occur in almost all common organic solvents, such as chloroform, toluene, tetrahydrofuran, dimethylformamide, and even in a mixed solvent of chloroform and methanol. However, they also found that, in pure chloroform, the assemblies were unstable and always fused into layers during solvent evaporation. Interestingly, the addition of methanol improved their stability, and very firm spheres were thus obtained from an optimized mixed solvent (volume ratio of chloroform:methanol, 4:1), which maintained a spherical shape quite stably even when free of solvent. A rapid assembly of $(DODMA)_4SiW_{12}O_{40}$ in mixed solvent (volume ratio of chloroform:methanol, 4:1) occurred in a very short time (0.5 min) after the sample was dissolved (Fig.3.32). The spherical assemblies are onionlike and exhibit ordinal circular multilamellae, with a layer spacing estimated at about 3.0 nm. And each fundamental layer has a sandwich structure in which alkyl chains shield the outside while POM clusters are located inside. The multilamellar HSEC spheres are morphologically similar to the onion phase of conventional surfactant vesicles, and empty spaces are found in the central part of some spheres, also like the microenvironment of vesicles formed by organic amphiphiles, which make these spherical assemblies suitable carriers to perform the catalytic and pharmacological functions of POMs.

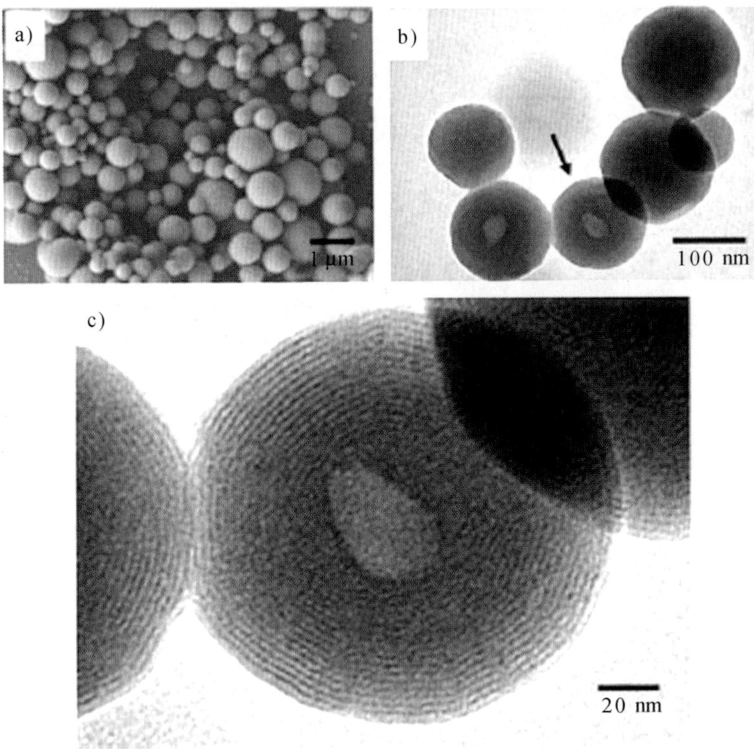

Fig.3.32. a) SEM and b) TEM images of $(DODMA)_4SiW_{12}O_{40}$ assemblies from a mixed solvent (chloroform:methanol, 4:1), and c) magnified image of the local region in b) indicated by an arrow. Reprinted from Li et al., 2007. Copyright (2007), with permission from Wiley

In addition to three-dimensional aggregates, one-dimensional nanowires incorporating POMs have also been reported. The mesoscopic organization of infinite $[Mo_3Se_3^-]_\infty$ chains in the presence of oppositely charged surfactants, dihexadecyldimethylammonium bromide (DDAB) and ω-undecenyltrimethylammonium bromide (ω-UTAB) was observed (Fig.3.33). This is a direct result of complex formation induced by the electrostatic interactions between the inorganic POMs and the surfactants (Messer et al., 2000). These studies indicated that $[Mo_3Se_3^-]_\infty$/DDAB and

$[Mo_3Se_3^-]_\infty/\omega$-UTAB complexes consist of well-aligned microscopic fiber bundles or sheets.

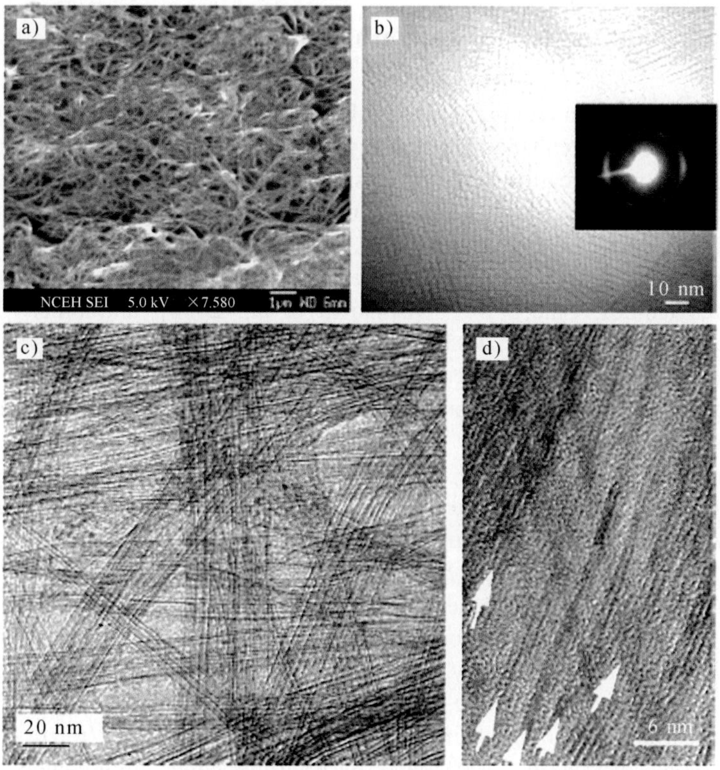

Fig.3.33. a) SEM image of $[Mo_3Se_3^-]_\infty$/DDAB complex nanofibers. b) TEM image of molecular wire $[Mo_3Se_3^-]_\infty$/DDAB complex bundles deposited on a TEM grid. Each dark line represents one individual $[Mo_3Se_3^-]_\infty$ wire. The inset shows an SAED pattern recorded on the sample area to its left. c) TEM image of $[Mo_3Se_3^-]_\infty/\omega$-UTAB complex bundles. d) High resolution TEM image of several $[Mo_3Se_3^-]_\infty/\omega$-UTAB complexed chains (indicated by arrows). Reprinted from Messer et al., 2000. Copyright (2000), with permission from Wiley

3.3 Self-assembled Honeycomb Films of Hydrophobic Surfactant-encapsulated Clusters (HSECs) at Air/Water Interface

From the above section, nano-scale POMs are able to interact with organic species, such as surfactants and polyelectrolytes, to form hydrophobic surfactant-encapsulated clusters (HSECs). We found a novel and interesting phenomenon for HSECs at the air/water interface, which are mainly discussed in this section.

3.3.1 Introduction to Honeycomb Films

Bees, the artificers in nature, always build their hives in a hexagonal arrangement to minimize work and enhance the stability of the hive. In the hive, each bee lives in one pore and is the regular neighbor of six others (Fig.3.34). In the laboratory, some researchers have successfully fabricated ordered porous films, apparently like natural beehives, here called "honeycomb films". The "replication" methods by homogeneous latex microsphere templates (Holland et al., 1998; Velev et al., 1999), emulsion templates (Imhof, Pine, 1998; Nishikawa et al., 2000), and other "lithograph" methods (Bolognesi et al., 2005), are still challenging due to the degree of order of templates and the removal of templates. In 1994, François and coworkers first found that a carbon disulphide solution of star-shaped polystyrene, which was cast on solid substrates with a moist airflow across the polymer solution surface to create a high humidity, could self assemble into an ordered pattern with hexagonally arranged pores 0.2~10 μm in diameter, walls 0.1~0.2 μm thick, and 10~30 μm high (Widawski et al., 1994). From then on, investigations of honeycomb films were mainly aimed at star-like polymers (Stenzel-Rosenbaum et al., 2001; Liu et al., 2007), block copolymers (Jenekhe, Chen, 1999; Zhao et al., 2003; Cheng et al., 2005), amphiphilic polyion complexes (Maruyama et al., 1998; Karthaus et al., 2000), and inorganic/organic hybrids (Karthaus et al., 1999) on solid substrates under humid conditions. Furthermore, the ordered self-assembly of honeycomb structures of fluorocarbon-stabilized silver nanoparticles at high humidity (relative humidity = 75%) (Yonezawa et al., 2001) and perfluoropolyether thiol-coated gold nanoparticles at 60% relative humidity (Shah et al., 2003) have also been fabricated by this method. However, no porous films can be obtained without high humidity. The ambient humidity around organic solutions is crucial to induce the self-assembly of materials into honeycomb films.

Very recently, it has been shown that surfactant-stabilized nano-scale POMs, which seem to be star-like spheres with POMs as the inner cores and surfactant chains outside, are able to self-assemble into ordered honeycomb films on solid substrates with a high humidity and/or simply at the air/water interface without any extra air flow. This is discussed in detail in the next subsection.

Ordered porous materials have been an attractive hotspot due to their potential applications in fields such as photonic and optoelectronic devices, membranes of separation, catalysis, and micrographic technology. Moreover, these honeycomb films also can be used as templates to replicate homogeneous micro-particles (De Boer et al., 2000) and microlenses (Yabu, Shimomura, 2005a), as cell culture substrates (Nishikawa et al., 1999), and as excellent superhydrophobic materials by using a fluorinated polymer (Yabu et al., 2005b).

Fig.3.34. Photograph of natural hive built by worker-bees

3.3.2 Fabricating Honeycomb Films of HSECs at Air/Water Interface

The hydrophobic surfactant-encapsulated clusters (HSECs) formed by electrostatic interaction between POMs and surfactants are able to self-assemble into ordered honeycomb films. To the best of our knowledge, reports on honeycomb films of POMs are very rare, since only two groups are working on them. Wu's group has reported a honeycomb film based on $(DODMA)_4H[Eu(H_2O)_2SiW_{11}O_{39}]$ complexes on solid substrates with a moist airflow across the solution surface (Bu et al., 2005). In our group, a simpler method is used. We successfully obtained self-assembled ordered honeycomb films of DODMA-encapsulated $\{Mo_{72}Fe_{30}\}$ and/or $\{Mn_2Bi_2W_{20}\}$ complexes at the air/water interface without any extra airflow (Fan et al., 2007; Tang, Hao, in preparation). In this condition, we believe that the humidity around the organic solution droplets is very high, nearly 100%. Taking DODMA-$\{Mo_{72}Fe_{30}\}$ complexes as an example, the honeycomb film at the air/water interface is fabricated as follows. For a 1.0 $mg \cdot mL^{-1}$ $\{Mo_{72}Fe_{30}\}$ aqueous solution mixed with enough DODMACl in $CHCl_3$, $\{Mo_{72}Fe_{30}\}$ can be transferred into the organic phase by forming hydrophobic $\{Mo_{72}Fe_{30}\}$/DODMA complexes through electrostatic interactions between negatively charged $\{Mo_{72}Fe_{30}\}$ anions and positively charged $DODMA^+$ cations. The yellow of the aqueous phase (upper phase) became colorless while the color of the $CHCl_3$-phase (lower phase) gradually turned yellow. Subsequently, an amount of $CHCl_3$-phase solution of $\{Mo_{72}Fe_{30}\}$/DODMA complexes was dropped onto a calm pure water surface. This initially formed a lens-shaped liquid film on the water surface. After the complete evaporation of $CHCl_3$, a thin opaque solid film remained on the water surface. The thin films can be transferred onto various solid substrates in order to characterize and further investigate the hydrophobic and electrochemical properties and the templates for electrochemical deposition.

We used a double long-chain surfactant dioctadecyldimethylammonium chloride (DODMACl), not single chain tetradecyltrimethylammonium bromide (TTABr), because DODMA can more effectively encapsulate $\{Mo_{72}Fe_{30}\}$ and the $\{Mo_{72}Fe_{30}\}$/DODMA complexes produced easily self-assemble into an ordered honeycomb film at the air/water interface. At $c_{DODMACl} = 2.0$ mg·mL^{-1}, TEM images of the fabricated thin film of $\{Mo_{72}Fe_{30}\}$/DODMA complexes at the air/water interface (Figs.3.35a and 3.35b) reveal the formation of thin films containing highly ordered, porous, honeycomb holes with a uniform size of about 3.5 μm. The average thickness of the walls between the pores remains at about 0.6 μm. HR-TEM studies show that the walls are composed of many 2.5 nm-diameter dark objects (Figs.3.35c and 3.35d) that can be attributed to single $\{Mo_{72}Fe_{30}\}$ clusters. This result confirms that the $\{Mo_{72}Fe_{30}\}$ clusters play a critical role in constructing the uniform honeycomb holes, together with surfactants. The highly ordered honeycomb architecture was also confirmed by SEM observations (Fig.3.36). From the SEM images, the holes are truly hollow and the architecture at the air/water interface has a thickness of about 500 nm.

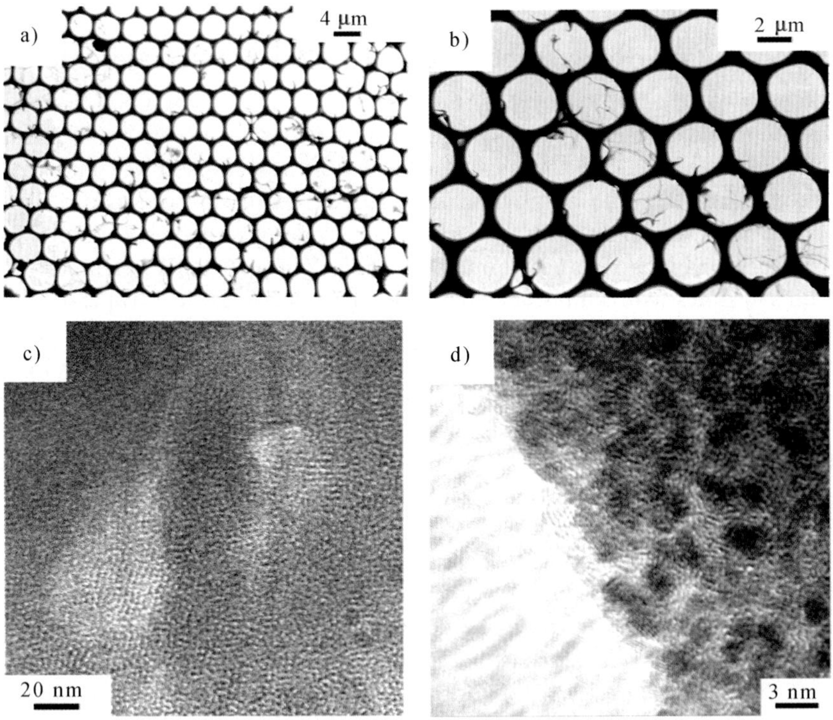

Fig.3.35. a), b) TEM images of highly ordered honeycomb architecture of $\{Mo_{72}Fe_{30}\}$/DODMA complexes at different magnification; and c), d) HR-TEM images of the highly ordered honeycombs revealing the ordered texture with $\{Mo_{72}Fe_{30}\}$ at $c_{DODMACl} = 2.0$ mg·mL^{-1} in $CHCl_3$. Reprinted from Fan et al., 2007. Copyright (2007), with permission from Wiley

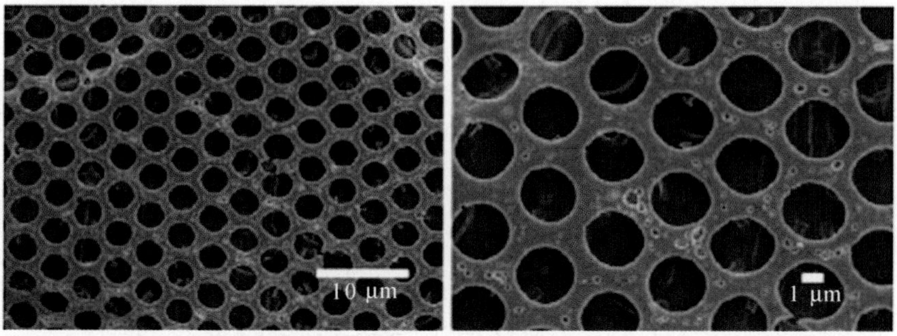

Fig.3.36. SEM images of highly ordered honeycomb architecture of $\{Mo_{72}Fe_{30}\}$/DODMA complexes at $c_{DODMACl}$ = 2.0 mg·mL^{-1} in $CHCl_3$. Reprinted from Fan et al., 2007. Copyright (2007), with permission from Wiley

In addition, sandwich-type $(DODMA)_{10}\{Mn_2Bi_2W_{20}\}$ complexes were directly dissolved in $CHCl_3$, and an ordered honeycomb film was fabricated using the same method. TEM image shows that the honeycomb film of $(DODMA)_{10}\{Mn_2Bi_2W_{20}\}$ complexes has a pore size of about 2 μm (Fig.3.37). XRD measurements (Fig.3.38) indicate that both the HSEC powder and the honeycomb film possess lamellar microstructures (Tang, Hao, 2009). Combining the lamellar distance of $DODMA^+$ (3.811 nm) (Okuyama et al., 1988) and the length of $\{Mn_2Bi_2W_{20}\}$ along the "*a*" direction (1.3 nm) (Bösing et al., 1998), the inferred interlamellar spacing of $(DODMA)_{10}\{Mn_2Bi_2W_{20}\}$ powders is consistent with the experimental value of 5.1 nm. The spacing of the honeycomb film of $(DODMA)_{10}\{Mn_2Bi_2W_{20}\}$ is shortened to 3.9 nm. Therefore, we speculate that $(DODMA)_{10}\{Mn_2Bi_2W_{20}\}$ clusters are packed much closer in the walls of the honeycomb films.

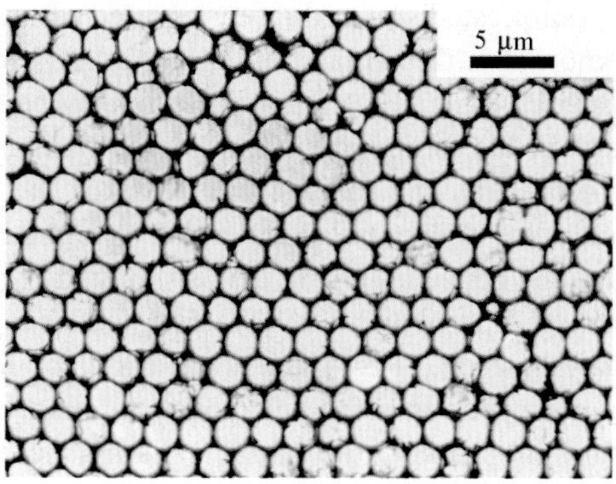

Fig.3.37. TEM image of thin film prepared by dropping 1.7 mg·mL^{-1} $(DODMA)_{10}\{Mn_2Bi_2W_{20}\}$ chloroform solution at the air/water interface. Reprinted from Tang, Hao, 2009. Copyright (2009), with permission from Elsevier

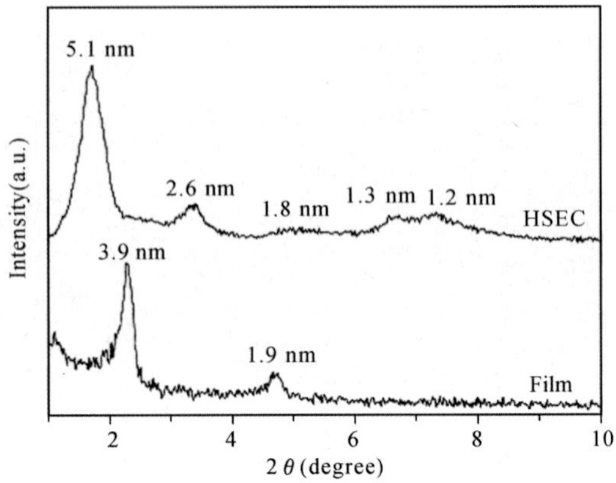

Fig.3.38. XRD patterns of HSEC powder and the corresponding honeycomb film of $(DODMA)_{10}\{Mn_2Bi_2W_{20}\}$. Reprinted from Tang, Hao, 2009. Copyright (2009), with permission from Elsevier

Based on many investigations of other POMs by our group, it can be concluded that, after long-chain surfactant encapsulation, it is usual for three-dimensional structural POMs to self-assemble into ordered honeycomb films at the air/water interface without any extra moist airflow.

3.3.3 Mechanism of Self-assembly of HSECs into Honeycomb Films

How do polymers and/or HSECs self-assemble into ordered honeycomb films on solid substrates with a moist airflow and/or directly at the air/water interface? We believe this is very complicated. According to reported results and our experimental investigations, we know it is the ambient humidity around organic solutions that is crucial to induce the self-assembly of materials into honeycomb films. Here, we propose a mechanism for the formation of honeycomb films in which the micrometer water droplets, condensed by the rapid evaporation of the organic solvent which cools the surface below the dew point of water, act as a good template for the pores, and the organic solutions concentrate and deposit around the water droplets (Pitois, François, 1999; Stenzel-Rosenbaum, 2002). The induced Marangoni convection (Block, 1956; Widawski et al., 1994) and local capillary forces mainly drive the arrangement of water droplets into the hexagonal minimum energy state. After the complete evaporation of solvent, the surface warms up to equilibrium, so the enveloped water droplets evaporate due to the increasing saturated vapor pressure. Finally, a solid thin film is obtained. In short, the stability of the water droplets which are the templates for honeycomb morphology is the crucial step (Fig.3.39). Therefore, influences can be used to control the energy state of water droplets, and consequently, the morphology of the hybrid films can be modulated.

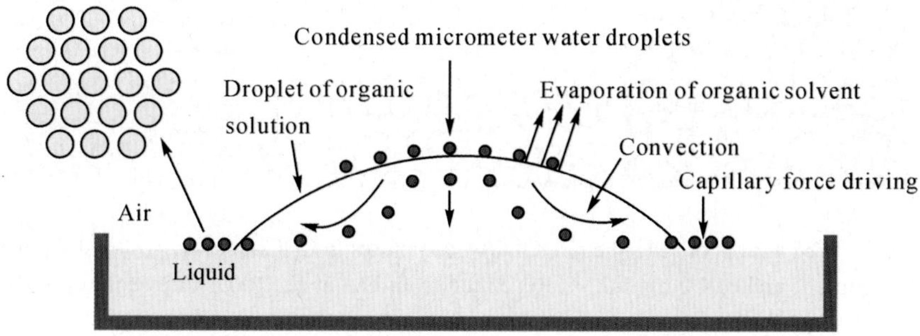

Fig.3.39. The formation mechanism during self-assembly of periodic hexagonal honeycomb films at the air/water interface. Reprinted from Tang, Hao, 2009. Copyright (2009), with permission from Elsevier

3.3.4 Morphology Modulation of Honeycomb Films of HSECs

During the self-assembly of HSECs at the air/water interface, the condensed water micro-droplets are good templates to fabricate honeycomb pores. Here, some of the influences of modulating water droplet state are studied, and finally, the honeycomb film morphology becomes changeable.

We originally speculated that the fully hydrophobic complexes would spread homogeneously on the surface of water, similar to the case of spreading a layer of oil at the air/water interface. However, in reality we found that the arrangement of complexes was very complicated and dependent on DODMACl concentration. For the self-assembly of $\{Mo_{72}Fe_{30}\}$/DODMA complexes at the air/water interface, at $c_{DODMACl}$ = 1.2 mg·mL^{-1} in $CHCl_3$, a micro-porous film was observed with a wall thickness between pores in general less than 0.5 μm, and non-homogenous pore sizes, ranging from 0.5 to 1.5 μm (Fig.3.40a). At $c_{DODMACl}$ = 2.0 mg·mL^{-1}, TEM images of $\{Mo_{72}Fe_{30}\}$/DODMA complexes (Fig.3.40b = Fig.3.35a) revealed the formation of thin films containing highly ordered, porous, honeycomb holes with a uniform size of about 3.5 μm. The average thickness of the walls between the pores remained about 0.6 μm. When the concentration of DODMACl was >4.0 mg·mL^{-1} in $CHCl_3$ (Fig.3.40c), TEM images showed that the unique honeycomb structures were destroyed. As we know, with increasing concentrations of DODMACl, some of the DODMACl is used to encapsulate POMs, and the excess is free in $CHCl_3$ which arranges at the water droplet/$CHCl_3$ solution droplet interface and strongly reduces the interfacial tension. An optimum surfactant concentration is fit for self-assembly into a honeycomb film. At lower or higher concentrations, no ordered honeycomb film is observed.

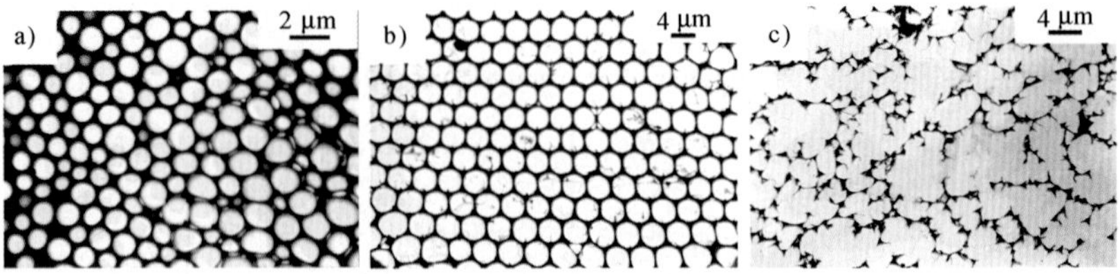

Fig.3.40. TEM images of $\{Mo_{72}Fe_{30}\}$/DODMA complexes in $CHCl_3$ at a) $c_{DODMACl} = 1.2\ mg \cdot mL^{-1}$, b) $2.0\ mg \cdot mL^{-1}$, and c) $4.0\ mg \cdot mL^{-1}$. Reprinted from Fan et al., 2007. Copyright (2007), with permission from Wiley

The rapid evaporation of solvents cools the water vapor into micrometer water droplets which are crucial templates for the formation of a honeycomb film. So the evaporation rate of solvents should also be considered (Tang, Hao, 2009). It was found that both carbon disulfide solution and chloroform solution of $(DODMA)_{10}\{Mn_2Bi_2W_{20}\}$ formed highly ordered honeycomb films at the air/water interface (Figs.3.41a and 3.41b) with a pore size of about 1 μm taking carbon disulfide as the solvent and a pore size of about 2 μm with chloroform. It is known that the boiling points of carbon disulfide and chloroform are 46 °C and 61 °C, respectively. This indicates that the pore size of the films becomes larger with the boiling point of the solvent. We speculate that, generally, when the boiling point increases, the volatility of the solvent decreases, so the condensed water droplets have more time to coalesce and grow during the self-organization; consequently, the pores templated by water droplets are much larger. However, if the boiling point of the solvent is very high and the volatility is very low, such as in cyclohexane and *n*-heptane with boiling points of 81 °C and 98 °C, respectively, no porous architecture is observed (Figs.3.41c and 3.41d), which may result from the very low volatility of cyclohexane and *n*-heptane. The solvent volatility is too slow to produce a sufficient temperature gradient around the organic solution surface, thus there are not enough condensed micrometer water droplets to template the formation of pores. Finally, some disordered fragments of HSECs are seen. So, it is concluded that solvents can not only modulate the pore size of the honeycomb film, but also determine whether or not a regular porous structure can be achieved at all.

Taking chloroform as a typical solvent, different HSECs including $(DODMA)_{10}\{Mn_2Bi_2W_{20}\}$, $(DDDMA)_{10}\{Mn_2Bi_2W_{20}\}$ and $(CTA)_{10}\{Mn_2Bi_2W_{20}\}$ can dissolve well (DODMA = dioctadecyldimethylammonium, DDDMA = didodecyldimethylammonium, CTA = cetyltrimethylammonium). The surface properties of HSECs, such as the hydrophobicity and the surface coverage, are clearly changed by encapsulation with different surfactants. We investigated the thin films of HSECs at the same amount of $\{Mn_2Bi_2W_{20}\}$, which were fabricated by casting the chloroform solutions of HSECs at the air/water interface (Fig.3.42). Both $(DODMA)_{10}\{Mn_2Bi_2W_{20}\}$ and $(DDDMA)_{10}\{Mn_2Bi_2W_{20}\}$ (but not $(CTA)_{10}\{Mn_2Bi_2W_{20}\}$) self-assembled into porous films at the air/water interface after rapid evaporation of chloroform. However, the arrangement of $(DDDMA)_{10}\{Mn_2Bi_2W_{20}\}$ film pores were much less ordered (Fig.3.42b). The frequencies

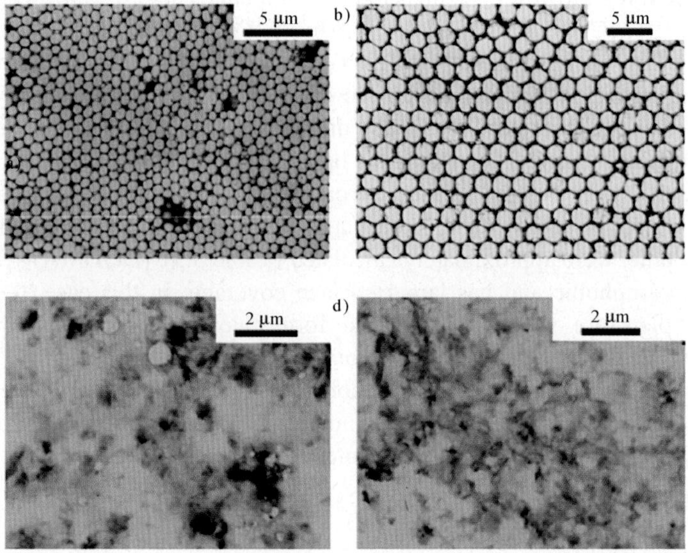

Fig.3.41. TEM images of thin films prepared by dropping 1.7 mg·mL^{-1} $(DODMA)_{10}\{Mn_2Bi_2W_{20}\}$ solution at the air/water interface using different solvents: a) carbon disulfide, b) chloroform, c) cyclohexane, and d) *n*-heptane. Reprinted from Tang, Hao, 2009. Copyright (2009), with permission from Elsevier

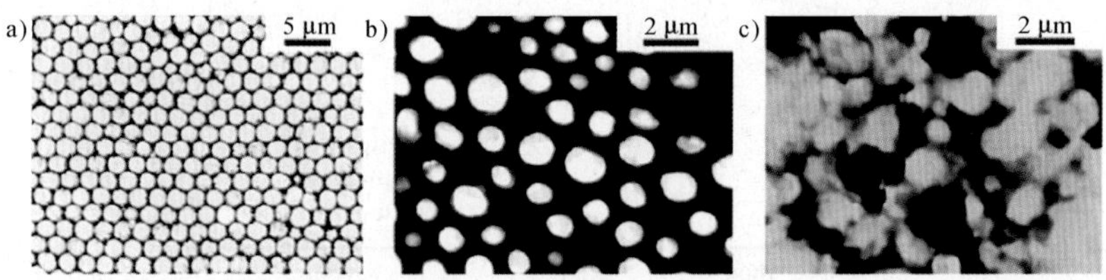

Fig.3.42. TEM images of thin films by casting three chloroform solutions at the air/water interface: a) 1.7 mg·mL^{-1} $(DODMA)_{10}\{Mn_2Bi_2W_{20}\}$; b) 1.5 mg·mL^{-1} $(DDDMA)_{10}\{Mn_2Bi_2W_{20}\}$; and c) 1.3 mg·mL^{-1} $(CTA)_{10}\{Mn_2Bi_2W_{20}\}$. Reprinted from Tang, Hao, 2009. Copyright (2009), with permission from Elsevier

for the CH_2 antisymmetric [ν_{as} (CH_2)] and the symmetric stretching [ν_s (CH_2)] bands are sensitive to the conformation of the alkyl chains (Nakashima et al., 1986). With the Fourier transform infrared (FT-IR) spectroscopy measurements of HSECs (Fig.3.43), the absorption bands at about 2,920 cm^{-1} [ν_{as} (CH_2)] and about 2,850 cm^{-1} [ν_s (CH_2)] attributed to the surfactant chains can show whether the hydrocarbon cationic surfactant is oriented on the $\{Mn_2Bi_2W_{20}\}^{10-}$ surface with winding or not. ν_{as} (CH_2) at 2,919 and 2,920 cm^{-1}, and ν_s (CH_2) at 2,849 and 2,850 cm^{-1} for $(DODMA)_{10}\{Mn_2Bi_2W_{20}\}$ and $(CTA)_{10}\{Mn_2Bi_2W_{20}\}$, respectively, indicate that the carbon chains are ordered on the $\{Mn_2Bi_2W_{20}\}^{10-}$ surface, which is crucial for high performance in construction of the honeycomb films. However, the higher absorption bands at 2,922 cm^{-1} [ν_{as} (CH_2)] and 2,851 cm^{-1} [ν_s (CH_2)] for $(DDDMA)_{10}$ $\{Mn_2Bi_2W_{20}\}$ indicate that $DDDMA^+$ chains are a little disordered on the $\{Mn_2Bi_2W_{20}\}^{10-}$ surface, which is a disadvantage for organizing an ordered structure. On the other hand, the hydrophobicity of HSECs mainly lies on the surfactant chains and the surface coverage of polyoxometalates by surfactant encapsulation. The hydrophilic lipophilic balance (HLB) values of surfactants (Zhao, Zhu, 2003) and the surface coverage of polyoxometalates by surfactant in HSECs, which is equal to the percentage of surfactant volume (Zana, 1980) to HSEC volume, were approximately calculated (Table 3.3). $(DODMA)_{10}\{Mn_2Bi_2W_{20}\}$ is more hydrophobic and has larger surface coverage; in this case, the condensed water droplets are well stabilized and form two-dimensional microcosmic w/o structures; finally, an ordered honeycomb film is self-organized at the air/water interface. However, the weak hydrophobicity and lower surface coverage cannot stabilize the water droplets well, so that no porous structure is obtained for $(CTA)_{10}\{Mn_2Bi_2W_{20}\}$ at the air/water interface. A higher hydrophobicity and larger surface coverage are propitious for the self-assembly of honeycomb films at the air/water interface.

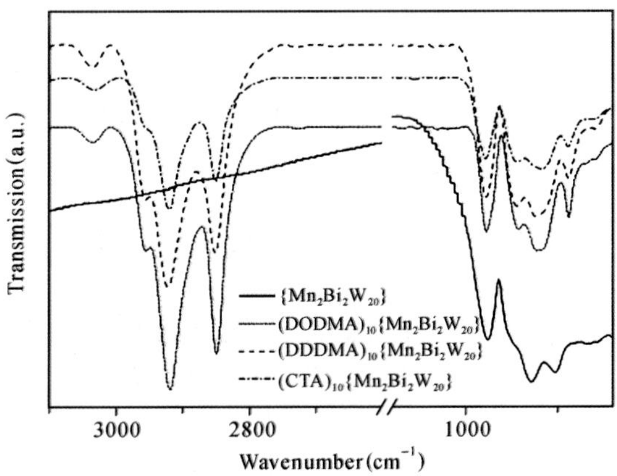

Fig.3.43. FT-IR spectroscopy of crystalline $\{Mn_2Bi_2W_{20}\}$ and powdered HSECs. Reprinted from Tang, Hao, 2009. Copyright (2009), with permission from Elsevier

Table 3.3. Properties of three cationic surfactants and the surface coverage of $\{Mn_2Bi_2W_{20}\}$ by surfactants. Reprinted from Tang, Hao, 2009. Copyright (2009), with permission from Elsevier

	Cationic surfactants		
	$DODMA^+$	$DDDMA^+$	CTA^+
HLB values	*a*–17.1	*a*–11.4	*a*–7.6
Volume of the hydrophobic chains nm^3	1.02	0.70	0.46
Surface coverage of POM in HSECs	79%	72%	63%

Note: "*a*" is a constant for the three homologous compounds

In addition, we found different morphologies of honeycomb films fabricated at different supporting surfaces. TEM images showed that thin films of $DODMA^+$-encapsulated $\{Mn_2Bi_2W_{20}\}$ at $c_{DODMA+} = 1.0\ mg\cdot mL^{-1}$ self-assembled at air/water, air/1 $mol\cdot L^{-1}$ NaCl solution, and air/2 $mmol\cdot L^{-1}$ TTABr solution interfaces (TTABr = tetradecyltrimethylammonium bromide) (Fig.3.44). At the air/water interface, the self-assembled honeycomb film had a few disordered pores (Fig.3.44a). When the surface tension of the supporting solution was increased by adding salts, such as at the air/1 $mol\cdot L^{-1}$ NaCl solution interface, the same chloroform solution self-assembled into a more closely-packed hexagonal porous film (Fig.3.44b). However, when a 2 $mmol\cdot L^{-1}$ TTABr solution acted as the supporting solution, because of amphiphilicity, the surfactant reduced the surface tension and caused a strong interfacial turbulence, so a thin film with very disordered fragments without pores occurred (Fig.3.44c).

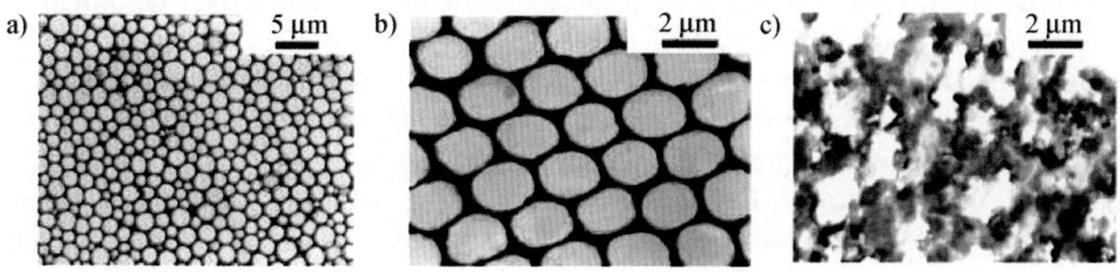

Fig.3.44. TEM images of thin films of $DODMA^+$-encapsulated $\{Mn_2Bi_2W_{20}\}$ at $c_{DODMA^+} = 1.0\ mg\cdot mL^{-1}$ fabricated at the a) air/water, b) air/1 $mol\cdot L^{-1}$ NaCl solution, and c) air/2 $mmol\cdot L^{-1}$ TTABr solution interfaces. Reprinted from Tang, Hao, 2009. Copyright (2009), with permission from Elsevier

3.4 Conclusions

Nano-scale polyoxometalates, as functional materials in a new century, are becoming attractive and widely investigated. These nano-clusters with beautiful topologies can be applied in catalysis, photonic-electronic devices, and medicine. Based on their good water solubility, according to electrostatic interactions, many organic molecules, including biological macromolecules, surfactants, and polyelectrolytes, can effectively transfer POMs into thin films to further construct functional interfacial materials by layer-by-layer and Langmuir-Blodgett techniques. Moreover, self-assembly of surfactant-encapsulated POM clusters into honeycomb films at the air/water interface, which is templated by condensed water micro-droplets are functionally attractive. There are still challenges and much more exploration is necessary in the development of POMs. So far, the valuable properties of POMs as catalysts and active antiviral agents have mainly been investigated on a few of the smaller heteropolyanions, while the mass of other POMs and the newly-synthesized giant POMs, such as $\{Mo_{368}\}$,

$\{Mo_{72}Fe_{30}\}$, and $\{Mn_2Bi_2W_{20}\}$, also need to be further studied. In addition, besides the basic investigations of the structure and assembly of POMs, organic-inorganic hybrids of multilayer films and porous films containing POMs are also worthy of thorough investigation for applications in catalysis, medicine, and material science. This will promote the development of interfacial films and build a bridge between polyoxometalate chemistry and material chemistry.

References

Ballardini R, Mulazzani QG, Venturi M, Bolletta F, Balzani V (1984) Photophysical characterization of the decatungstoeuropate (9-) anion. Inorg Chem 23:300-305

Bazzan G, Smith W, Francesconi LC, Drain CM (2008) Electrostatic self-organization of robust porphyrin-polyoxometalate films. Langmuir 24:3244-3249

Block MJ (1956) Surface tension as the cause of Benard cells and surface deformation in a liquid film. Nature 178:650-651

Bolognesi A, Mercogliano C, Yunus S, Civardi M, Comoretto D, Turturro A (2005) Self-organization of polystyrenes into ordered microstructured films and their replication by soft lithography. Langmuir 21:3480-3485

Bösing M, Nöh A, Loose I, Krebs B (1998) Highly efficient catalysts in directed oxygen-transfer processes: Synthesis structures of novel manganese-containing heteropolyanions and applications in regioselective epoxidation of dienes with hydrogen peroxide. J Am Chem Soc 120:7252-7259

Brown GM. Noe-Spirlet MR. Busing WR. Levy HA (1977) Dodecatungstophosphoric acid hexahydrate $(H_5O_2^+)_3(PW_{12}O_{40}^{3-})$: the true structure of Keggin's 'pentahydrate' from single-crystal X-ray and neutron diffraction data. Acta Crystallogr Sect B: Struct Sci 33:1038-1046

Bu W, Li H, Sun H, Yin S, Wu L (2005) Polyoxometalate-based vesicle and its honeycomb architectures on solid surfaces. J Am Chem Soc 127:8016-8017

Bu W, Wu L, Hou X, Fan H, Hu C, Zhang X (2002) Investigation on solvent casting films of surfactant-encapsulated clusters. J Colloid Interface Sci 251:120-124

Caruso F (2000) Hollow capsule processing through colloidal templating and self-assembly. Chem Eur J 6:413-419

Caruso F, Caruso RA, Möhwald H (1998c) Nanoengineering of inorganic and hybrid hollow spheres by colloidal templating. Science 282:1111-1114

Caruso F, Kurth DG, Volkmer D, Koop MJ, Müller A (1998a) Ultrathin molybdenum polyoxometalate-polyelectrolyte multilayer films. Langmuir 14:3462-3465

Caruso F, Lichtenfeld H, Giersig M, Möhwald H (1998b) Electrostatic self-assembly of silica nanoparticle-polyelectrolyte multilayers on polystyrene latex particles. J Am Chem Soc 120:8523-8524

Casan-Pastor N, Gómez-Romero P (2004) Polyoxometalates: from inorganic chemistry to materials science. Frontiers in Bioscience 9:1759-1770

Chen Z, Loo BH, Ma Y, Cao Y, He T, Yang W, Yao J (2004) Synthesis of novel photochromic material in the violet color region: composite 1-hexadecylammonium-polyoxomolybdate thin films. Mater Res Bull 39:1167-1173

Chen Z, Ma Y, Zhang X, Liu B, Yao J (2001) Photochromic inorganic/organic self-assembled superlattice films. J Colloid Interface Sci 240:487-491

Cheng CX, Tian Y, Shi YQ, Tang RP, Xi F (2005) Porous polymer films and honeycomb structures based on amphiphilic dendronized block copolymers. Langmuir 21:6576-6581

Clemente-León M, Agricole B, Mingotaud C, Gómez-Garcia CJ, Coronado E, Delhaes P (1997a) Toward new organic/inorganic superlattices: Keggin polyoxometalates in Langmuir and Langmuir-Blodgett films. Langmuir 13:2340-2347

Clemente-León M, Mingotaud C, Agricole B, Gómez-Garcia CJ, Coronado E, Delhaès P (1997b) Application of the Langmuir-Blodgett technique to polyoxometalates: towards new magnetic films. Angew Chem Int Ed Engl 36:1114-1116

Clemente-León M, Mingotaud C, Gomez-Garcia CJ, Coronado E, Delhaes P (1998a) Polyoxometalates in Langmuir-Blodgett films: toward new magnetic materials. Thin solid films 327:439-442

Clemente-León M, Soyer H, Coronado E, Mingotaud C, Gómez-García CJ, Delhaès P (1998b) Langmuir-Blodgett films of single-molecule nanomagnets. Angew Chem Int Ed 37:2842-2845

Coronado E, Gomez-Garcia CJ (1998) Polyoxometalate-based molecular materials. Chem Rev 98:273-296

Coronado E, Mingotaud C (1999) Hybrid organic/inorganic Langmuir-Blodgett films: a supramolecular approach to ultrathin magnetic films. Adv Mater 11:869-872

Cui J, Fan D, Hao J (2009) Magnetic $\{Mo_{72}Fe_{30}\}$-embedded hybrid nanocapsules. J Colloid Interface Sci 330:488-492

De Boer B, Stalmach U, Nijland H, Hadziioannou G (2000) Microporous honeycomb-structured films of semiconducting block copolymers and their use as patterned templates. Adv Mater 12:1581-1583

Decher G (1997) Fuzzy nanoassemblies: toward layered polymeric multicomposites Science 277:1232-1237

Donath E, Sukhorukov GB, Caruso F, Davis SA, Möhwald H (1998) Novel hollow polymer shells by colloid-templated assembly of polyelectrolytes. Angew Chem Int Ed 37:2201-2205

Fan D, Hao J (2009b) Fabrication and electrocatalytic properties of chitosan and Keplerate-type polyoxometalate $\{Mo_{72}Fe_{30}\}$ hybrid films. J Phys Chem B 113:7513- 7516

Fan D, Hao J (2009a) Phase stability of Keplerate-type polyoxomolybdates controlled by added cationic surfactant. J Colloid Interface Sci 333:757-763

Fan D, Jia X, Tang P, Hao J, Liu T (2007) Self-patterning of hydrophobic materials into highly ordered honeycomb nanostructures at the air/water interface. Angew Chem Int Ed 46:3342-3345

Gao G. Xu L. Wang W. An W. Qiu Y. Wang Z. Wang E (2005a) Electrochromic multilayer films of tunable color by combination of copper or iron complex and monolacunary Dawson-type polyoxometalate. J Phys Chem B 109:8948-8953

Gao L. Wang E. Kang Z. Song Y. Mao B. Xu L (2005b) Layer-by-layer assembly of polyoxometalates into microcapsules. J Phys Chem B 109:16587-16592

Gaponik N. Radtchenko IL. Sukhorukov GB. Rogach AL (2004) Luminescent polymer microcapsules addressable by a magnetic field. Langmuir 20:1449-1452

Holland BT, Blanford CF, Stein A (1998) Synthesis of macroporous minerals with highly ordered three-dimensional arrays of spheroidal voids. Science 281:538-540

Imhof A, Pine DJ (1998) Uniform macroporous ceramics and plastics by emulsion templating. Adv Mater 10:697-700

Jabbour D. Keita B. Nadjo L. Kortz U. Mal SS (2005) The wheel-shaped Cu_{20}-tungstophosphate $[Cu_{20}Cl(OH)_{24}(H_2O)_{12}(P_8W_{48}O_{184})]^{25-}$ redox and electrocatalytic properties. Electrochem Commun 7:841-847

Jenekhe SA. Chen XL (1999) Self-assembly of ordered microporous materials from rod-coil block copolymers. Science 283:372-375

Jiang M, Wang E, Kang Z, Lian S, Wu A, Li Z (2003) In situ controllable synthesis of polyoxometalate nanoparticles in polyelectrolyte multilayers. J Mater Chem 13: 647-649

Jiang M, Zhai X, Liu M (2005) Fabrication and photoluminescence of hybrid organized molecular films of a series of gemini amphiphiles and europium (III)-containing polyoxometalate. Langmuir 21:11128-11135

Karthaus O, Cieren X, Maruyama N, Shimomura M (1999) Mesoscopic 2-D ordering of inorganic/organic hybrid materials. Mater Sci Eng C 10:103-106

Karthaus O, Maruyama N, Cieren X, Shimomura M, Hasegawa H, Hashimoto T (2000) Water-assisted formation of micrometer-size honeycomb patterns of polymers. Langmuir 16:6071-6076

Katsoulis DE (1998) A survey of applications of polyoxometalates. Chem Rev 98:359-387

Keita B. Zhang G. Dolbecq A. Mialane P. Sécheresse F. Miserque F. Nadjo L (2007) Mo^{V}-Mo^{VI} mixed valence polyoxometalates for facile synthesis of stabilized metal nanoparticles: electrocatalytic oxidation of alcohols. J Phys Chem C 111:8145-8148

Kortz U, Jeannin YP, Tézé A, Hervé G, Isber SA (1999) Novel dimeric Ni-substituted β-keggin silicotungstate: structure and magnetic properties of $K_{12}[\beta\text{-}SiNi_2W_{10}O_{36}(OH)_2(H_2O)\}_2]\cdot 20H_2O$. Inorg Chem 38:3670-3675

Kortz U, Müller A, Slageren J, Schnack J, Dalal NS, Dressel M (2009) Polyoxometalates: fascinating structures unique magnetic properties. Coord Chem Rev In Press

Kozhevnikov IV (1998) Catalysis by heteropoly acids and multicomponent polyoxometalates in liquid-phase reactions. Chem Rev 98:171-198

Kurth DG, Lehmann P, Volkmer D, Cöelfen H, Koop MJ, Müller A, Du Chesne A (2000a) Surfactant-encapsulated clusters (SECs): $(DODA)_{20}(NH_4)[H_3Mo_{57}V_6(NO)_6O_{183}(H_2O)_{18}]$, a case study. Chem Eur J 6:385-393

Kurth DG, Lehmann P, Volkmer D, Müller A, Schwahn D (2000c) Biologically inspired polyoxometalate-surfactant composite materials investigations on the structures of discrete surfactant-encapsulated clusters monolayers and Langmuir-Blodgett films of $(DODA)_{40}(NH_4)_2[(H_2O)_nMo_{132}O_{372}(CH_3CO_2)_{30}(H_2O)_{72}]$. J Chem Soc Dalton Trans 3989: 3989-3998

Kurth DG, Volkmer D, Ruttorf M, Richter B, Müller A (2000b) Ultrathin composite films incorporating the nanoporous isopolyoxomolybdate "keplerate" $(NH_4)_{42}[Mo_{132}O_{372}(CH_3COO)_{30}(H_2O)_{72}]$. Chem Mater 12:2829-2831

Lee IS, Long JR, Prusiner SB, Safar JG (2005) Selective precipitation of prions by polyoxometalate complexes. J Am Chem Soc 127:13802-13803

Leporatti S, Voigt A, Mitlöhner R, Sukhorukov G, Donath E, Möhwald H (2000) Scanning force microscopy investigation of polyelectrolyte nano- and microcapsule wall texture. Langmuir 16:4059-4063

Li H, Sun H, Qi W, Xu M, Wu L (2007) Onionlike hybrid assemblies based on

surfactant-encapsulated polyoxometalates. Angew Chem Int Ed 46:1300-1303
Li W, Yin S, Wang J, Wu L (2008) Tuning mesophase of ammonium amphiphile-encapsulated polyoxometalate complexes through changing component structure. Chem Mater 20:514-522
Liu C, Gao C, Yan D (2007) Honeycomb-patterned photoluminescent films fabricated by self-assembly of hyperbranched polymers. Angew Chem Int Ed 46:4128-4131
Liu S, Kurth DG, Bredenkötter B, Volkmer D (2002a) The structure of self-assembled multilayers with polyoxometalate nanoclusters. J Am Chem Soc 124:12279-12287
Liu S, Kurth DG, Möhwald H, Volkmer D (2002b) A thin-film electrochromic device based on a polyoxometalate cluster. Adv Mater 14:225-228
Liu S, Möhwald H, Volkmer D, Kurth DG (2006b) Polyoxometalate-based electro- and photochromic dual-mode devices. Langmuir 22:1949-1951
Liu S, Tang Z, Wang E, Dong S (1999) Fabrication and characterization of heteropolyanion Langmuir-Blodgett films. Thin Solid Films 339:277-283
Liu S, Volkmer D, Kurth DG (2003) Functional polyoxometalate thin films via electrostatic layer-by-layer self-assembly. J Cluster Sci 14:405-419
Liu T (2003) Surfactant-induced trans-interface transportation and complex formation of giant polyoxomolybdate-based clusters. J Cluster Sci 14:215-226
Liu T, Imber B, Diemann E, Liu G, Cokleski K, Li H, Chen Z, Müller A (2006a) Deprotonations and charges of well-defined $\{Mo_{72}Fe_{30}\}$ nanoacids simply stepwise tuned by pH allow control/variation of related self-assembly processes. J Am Chem Soc 128:15914-15920
López X, Maestre JM, Bo C, Poblet JM (2001) Electronic properties of polyoxometalates: a DFT study of α/β-$[XM_{12}O_{40}]^{n-}$ relative stability (M= W Mo and X a main group element). J Am Chem Soc 123:9571-9576
Maruyama N, Koito T, Nishida J, Sawadaishi T, Cieren X, Ijiro K, Karthaus O, Shimomura M (1998) Mesoscopic patterns of molecular aggregates on solid substrates. Thin Solid Films 327-329:854-856
Mbomekalle IM, Keita B, Lu YW, Nadjo L, Contant R, Belai N, Pope MT (2004) Synthesis characterization and electrochemistry of the novel Dawson-type tungstophosphate $[H_4PW_{18}O_{62}]^{7-}$ and first transition metal ions derivatives. Eur J Inorg Chem 2:276-285
Messer B, Song JH, Huang M, Wu Y, Kim F, Yang P (2000) Surfactant-induced mesoscopic assemblies of inorganic molecular chains. Adv Mater 12:1526-1528
Moriguchi I, Fendler JH (1998) Characterization and electrochromic properties of ultrathin films self-assembled from poly (diallyldimethylammonium) chloride and sodium decatungstate. Chem Mater 10:2205-2211
Müller A, Beckmann E, Bögge H, Schmidtmann M, Dress A (2002) Inorganic chemistry goes protein size: a Mo_{368} nano-hedgehog initiating nanochemistry by symmetry breaking. Angew Chem Int Ed 41:1162-1167
Müller A, Botar B, Das SK, Bögge H, Schmidtmann M, Merca A (2004) On the complex hedgehog-shaped cluster species containing 368 Mo atoms: simple preparation method new spectral details and information about the unique formation. Polyhedron 23:2381-2385
Müller A, Das SK, Fedin VP, Krickemeyer E, Beugholt C, Bögge H, Schmidtmann M, Hauptfleisch B (1999b) Rapid and simple isolation of the crystalline molybdenum-blue compounds with discrete and linked nanosized ring-shaped anions. Z Anorg

Allg Chem 625:1187-1192

Müller A, Kögerler P, Kuhlmann C (1999c) A variety of combinatorially linkable units as disposition: from a giant icosahedral Keplerate to multi-functional metal-oxide based network structures. Chem Commun 1999:1347-1358

Müller A, Krickemeyer E, Bögge H, Schmidtmann M, Peters F (1998b) Organizational forms of matter: an inorganic super fullerene and keplerate based on molybdenum oxide. Angew Chem Int Ed 37:3360-3364

Müller A, Luban M, Schröder C, Modler R, Kögerler P, Axenovich M, Schnack J, Canfield P, Bud'ko S, Harrison N (2001) Classical and quantum magnetism in giant keplerate magnetic molecules. ChemPhysChem 2:517-521

Müller A, Meyer J, Krickemeyer E, Beugholt C, Bögge H, Peters F, Schmidtmann M, Kögerler P, Koop MJ (1998a) Unusual stepwise assembly and molecular growth: $[H_{14}Mo_{37}O_{112}]^{14-}$ and $[H_3Mo_{57}V_6(NO)_6O_{189}(H_2O)_{12}(MoO)_6]^{21-}$. Chem Eur J 4:1000-1006

Müller A, Peters F, Pope MT, Gatteschi D (1998c) Polyoxometalates: very large clusters-nanoscale magnets. Chem Rev 98:239-272

Müller A, Sarkar S, Shah SQN, Bögge H, Schmidtmann M, Sarkar S, Kögerler P, Hauptfleisch B, Trautwein AX, Schünemann V (1999a) Archimedean synthesis and magic numbers: “sizing” giant molybdenum-oxide-based molecular spheres of the keplerate type. Angew Chem Int Ed 38:3238-3241

Müller A, Serain C (2000) Soluble molybdenum blues “des pudels kern”. Acc Chem Res 33:2-10

Nakashima N, Yamada N, Kunitake T (1986) A Fourier transform infrared study of bilayer membranes of double-chain ammonium amphiphiles. J Phys Chem 90:3374-3377

Nishikawa T, Nishida J, Ookura R, Nishimura SI, Scheumann V, Zizlsperger M, Lawall R, Knoll W, Shimomura M (2000) Web-structured films of an amphiphilic polymer from water in oil emulsion: fabrication and characterization. Langmuir 16:1337-1342

Nishikawa T, Nishida J, Ookura R, Nishimura SI, Wada S, Karino T, Shimomura M (1999) Honeycomb-patterned thin films of amphiphilic polymers as cell culture substrates. Mater Sci Eng C 8-9:495-500

Okuyama K, Soboi Y, Iijima N, Hirabayashi K, Kunitake T, Kajiyama T (1988) Molecular and crystal structure of the lipid-model amphiphile dioctadecyldimethylammonium bromide monohydrate. Bull Chem Soc Jpn 61:1485-1490

Pichon C, Mialane P, Dolbecq A, Marrot J, Rivière E, Bassil BS, Kortz U, Keita B, Nadjo L, Sécheresse F (2008) Octa-and nonanuclear nickel (II) polyoxometalate clusters: synthesis and electrochemical and magnetic characterizations. Inorg Chem 47:11120-11128

Pitois O, François B (1999) Crystallization of condensation droplets on a liquid surface. Colloid Polym Sci 277:574-578

Polarz S, Smarsly B, Antonietti M (2001) Colloidal organization and clusters: self-assembly of polyoxometalate-surfactant complexes towards three-dimensional organized structures. ChemPhysChem 2:457-461

Pope MT, Müller A (1991) Polyoxometalate chemistry: an old field with new dimensions in several disciplines. Angew Chem Int Ed 30:34-48

Prinz M, Takács AF, Schnack J, Balasz I, Burzo E, Kortz U, Kuepper K, Neumann M

(2006) Magnetic and electronic properties of the iron-containing polyoxotungstate $(Fe_4(H_2O)_{10}(\beta\text{-}SbW_9O_{33})_2)^{6-}$. J Appl Phys 99:08J505-1-3

Proust A, Thouvenot R, Gouzerh P (2008) Functionalization of polyoxometalates: towards advanced applications in catalysis and materials science. Chem Commun 2008:1837-1852

Raynaud M, Chermann JC, Plata F, Jasmin C, Mathé G (1971) Viral inhibitors of murine leukemia-sarcoma group Tungstosilicate. C R Acad Sci Ser D 272:347-348

Rhule JT, Hill CL, Judd DA, Schinazi RF (1998) Polyoxometalates in medicine. Chem Rev 98:327-358

Sadakane M, Steckhan E (1998) Electrochemical properties of polyoxometalates as electrocatalysts. Chem Rev 98:219-237

Secheresse F, Ratajczak H, Müller A (2002) Foreword. J Cluster Sci 13:265-266

Shah PS, Sigman MB, Stowell CA, Lim KT, Johnston KP, Korgel BA (2003) Single-step self-organization of ordered macroporous nanocrystal thin films. Adv Mater 15:971-974

Shchukin DG, Shutava T, Shchukina E, Sukhorukov GB, Lvov YM (2004) Modified polyelectrolyte microcapsules as smart defense systems. Chem Mater 16:3446-3451

Shchukin DG, Sukhorukov GB, Möhwald H (2003b) Smart inorganic/organic nanocomposite hollow microcapsules. Angew Chem Int Ed 42:4472-4475

Shchukin DG, Ustinovich E, Sviridov DV, Lvov YM, Sukhorukov GB (2003a) Photocatalytic microreactors based on TiO_2-modified polyelectrolyte multilayer capsules. Photochem Photobiol Sci 2:975-977

Shchukin DG, Ustinovich EA, Sukhorukov GB, Möhwald H, Sviridov DV (2005) Metallized polyelectrolyte microcapsules. Adv Mater 17:468-472

Stenzel-Rosenbaum MH (2002) Formation of regular honeycomb-patterned porous film by self-organization. Aust J Chem 55:239-243

Stenzel-Rosenbaum MH, Davis TP, Fane AG, Chen V (2001) Porous polymer films and honeycomb structure made by the self-organization of well-defined macromolecular structures created by living radical polymerization techniques. Angew Chem Int Ed 40:3428-3432

Sugeta M, Yamase T (1993) Crystal structure and luminescence site of $Na_9[EuW_{10}O_{36}]\cdot 32H_2O$. Bull Chem Soc Jpn 66:444-449

Sukhorukov GB, Donath E, Lichtenfeld H, Knippel E, Knippel M, Budde A, Möhwald H (1998) Layer-by-layer self assembly of polyelectrolytes on colloidal particles. Colloids Surf A 137:253-266

Sukhorukov GB, Volodkin DV, Günther AM, Petrov AI, Shenoy DB, Möhwald H (2004) Porous calcium carbonate microparticles as templates for encapsulation of bioactive compounds. J Mater Chem 14:2073-2081

Sun H, Li H, Bu W, Xu M, Wu L (2006) Self-organized microporous structures based on surfactant-encapsulated polyoxometalate complexes. J Phys Chem B 110:24847-24854

Tang P, Hao J (2009) Formation mechanism and morphology modulation of honeycomb hybrid films made of polyoxometalates/surfactants at the air/water interface. J Colloid Interface Sci 333:1-5

Tang P, Hao J (In prepared) Universal self-patterning of honeycomb-structured microporous films using nano-metre sized inorganic clusters encapsulated with surfactants at the air/water interface.

Tang P, Jia X, Fan D, Wang L, Hao J (2008) Surface charges of hedgehog-shaped

polyoxomolybdate modified by a cationic surfactant and the inorganic/organic complex. Colloids Surf A 312:18-23
Velev OD, Tessier PM, Lenhoff AM, Kaler EW (1999) A class of porous metallic nanostructures. Nature 401:548-548
Volkmer D, Bredenkötter B, Tellenbröker J, Kögerler P, Kurth DG, Lehmann P, Schnablegger H, Schwahn D, Piepenbrink M, Krebs B (2002) Structure and properties of the Dendron-encapsulated polyoxometalate $(C_{52}H_{60}NO_{12})_{12}[(Mn(H_2O))_3(SbW_9O_{33})_2]$, a first generation dendrizyme. J Am Chem Soc 124:10489-10496
Volkmer D, Du Chesne A, Kurth DG, Schnablegger H, Lehmann P, Koop MJ, Müller A (2000) Toward nanodevices: synthesis and characterization of the nanoporous surfactant-encapsulated Keplerate $(DODA)_{40}(NH_4)_2[(H_2O)_nMo_{132}O_{372}(CH_3COO)_{30}(H_2O)_{72}]$. J Am Chem Soc 122:1995-1998
Wang XL, Qin C, Wang EB, Su ZM, Li YG, Xu L (2006) Self-assembly of nanometer-scale $[Cu_{24}I_{10}L_{12}]^{14+}$ cages and ball-shaped Keggin clusters into a (4,12)-connected 3D framework with photoluminescent and electrochemical properties. Angew Chem Int Ed 45:7411-7414
Widawski G, Rawiso M, François B (1994) Self-organized honeycomb morphology of star-polymer polystyrene films. Nature 369:387-389
Xu L, Zhang H, Wang E, Kurth DG, Li Z (2002) Photoluminescent multilayer films based on polyoxometalates. J Mater Chem 12:654-657
Yabu H, Shimomura M (2005a) Simple fabrication of micro lens arrays. Langmuir 21:1709-1711
Yabu H, Takebayashi M, Tanaka M, Shimomura M (2005b) Superhydrophobic and lipophobic properties of self-organized honeycomb and pincushion structures. Langmuir 21:3235-3237
Yin S, Sun H, Yan Y, Li W, Wu L (2009) Hydrogen-bonding-induced supramolecular liquid crystals and luminescent properties of europium-substituted polyoxometalate hybrids. J Phys Chem B 113:2355-2364
Yonezawa T, Onoue S, Kimizuka N (2001) Self-organized superstructures of fluorocarbon-stabilized silver nanoparticles. Adv Mater 13:140-142
Zana R (1980) Ionization of cationic micelles: effect of the detergent structure. J Colloid Interface Sci 78:330-337
Zhao G, Zhu B (2003) Principles of surfactant action. Beijing: China Light Industry Press
Zhao X, Cai Q, Shi G, Shi Y, Chen G (2003) Formation of ordered microporous films with water as templates from poly (D L-lactic-co-glycolic acid) solution. J Appl Polym Sci 90:1846-1850

4

Natural Cellulosic Substance Derived Nano-structured Materials

Yuanqing Gu, Jianguo Huang*
*Department of Chemistry, Zhejiang University, Hangzhou, Zhejiang, 310027, China.
E-mail: jghuang@zju.edu.cn

When versatile synthetic chemical processes meet natural biological assemblies, a promising shortcut for the design and fabrication of functional materials with tailored structures and properties are lit up. By precisely replicating natural substrates with guest matrices, artificial materials are endowed with the initial biological structures and morphologies. To achieve faithful inorganic/organic replicas of the natural species for the corresponding finest structural details and morphological hierarchies, one effective and practical strategy is to coat the morphologically sophisticated surfaces of the biological structures with ultrathin films accompanied by subsequent removal of the biotemplate. With this process, the morphological hierarchies of initial biological substances can be replicated faithfully from macroscopic down to nanometer scales. And it was successfully applied to natural cellulosic substances such as filter paper, cotton, and cloth to yield the related metal oxide replicas. The hierarchical structure and highly detailed morphologies of the cellulosic substances are precisely memorized in metal oxide films to give macroscopic fossils; and the organic substances are removed by subsequent calcination. The resultant fossils are hierarchical ceramic materials, in which the structures of the original template substance are faithfully inherited. The ceramics are composed of metal oxide nanotubes, as precise hollow replicas of the template cellulose nanofibers. This approach has been employed to synthesize titania, zirconia, tin oxide, and ITO nanotubular materials. Hierarchical titania nanotube-gold nanoparticle hybrid and polypyrrole composite materials are also achieved with using filter paper as a scaffold. Also, the titania-coated cellulose fibers are employed as a substrate for protein immobilization, resulting in novel bioactive materials. Furthermore, by dissolving the cellulose template instead of calcination, this approach is extended to the design and preparation of bio-inspired polymeric nanotubular materials.

4.1 Introduction

The distinguished and astonishing properties of natural substances can be originated from their unique nature-produced hierarchical structures. Therefore, replication of the unique and complicated multilevel morphologies and structures of biological template substrates, with an inorganic or organic matrix, is believed to be able to introduce some of the superb properties of biological organisms into artificial materials. That is, replication with natural substances as templates provides a facile, low-cost, and environmentally benign pathway for design and fabrication of advanced functional materials (Sanchezl et al., 2005; Pouget et al., 2007).

Indeed, a large variety of biological species including bacterium (Davis et al., 1997), diatom (Anderson et al., 2000), living cells (Chia et al., 2000), skeleton (Meldrum, Seshadri, 2000), eggshell membrane (Yang et al., 2002), wood (Shin et al., 2001; Dong et al., 2002), silk (Kim, 2003), pollen (Hall et al., 2003), and butterfly wing (Huang et al., 2006), for example, have been employed as initial biotemplates. Analogous materials were synthesized in the forms of negative/ positive or true copies with different chemical processes including chemical vapor deposition (CVD) (Cook et al., 2003), atomic layer deposition (ALD) (Kemell et al., 2005), gas-solid displacement reactions (Bao et al., 2007), as well as wet chemistry techniques like sol–gel polycondensation (Caruso, Antonietti, 2001). The majority of the biotemplates show unique structures such as nanoporous features, channels, and other complex hierarchical architectures. Unfortunately, in these examples, the morphological replication is generally precise only down to the micrometer scale, and the nanoscopic details are failed to be reproduced. Thus, the fine structural organizations of the original biological templates are not able to be faithfully inherited by the resultant replicas; and further properties related to the unique morphological features of the natural substances are not maintained.

Therefore, it still remains a challenge to achieve analogues that faithfully inherit the corresponding finest structural details and morphological hierarchies of templates from natural biological species at nanoprecision. Siliceous woods, an independently existing silica analogy of fossils formed from the original plants with corresponding intricate details and hierarchical structures, demonstrates a short-cut route to reproduce the original structures and morphologies. Such fossilization processes inspire material researchers to realize that artificial replication of morphologically sophisticated surfaces of biological structures is possible by forming ultrathin films faithfully lined over the biological templates, accompanied by subsequent removal of the original substrates. To this end, the surface sol–gel strategy was developed as a powerful tool for faithful replication of the hierarchical structures and characteristic morphologies of natural substances from the macroscopic down to the nanolevel details. Till now, this methodology was successfully applied to natural cellulosic substances such as filter paper, cotton, and cloth to design and fabricate various related functional metal oxide materials like titania (Huang, Kunitake, 2003; Caruso, 2004), zirconia (Huang, Kunitake, 2003; Caruso, 2004), tin oxide (Huang et al., 2005), and indium tin oxide (Aoki et al., 2006).

Besides the inorganic materials obtained using metal oxides, guest substrates, such as a conjugated conducting polymer (*e.g.*, polypyrrole), were also employed to coat each nanofiber surface of the cellulosic substances, like filter paper, with nanometer precision, using the polymerization-induced adsorption approach to produce nanostructured, porous polymer materials (Huang et al., 2005). Furthermore, the aforementioned extremely thin metal oxide films deposited over the surfaces of the biological substances can act as a platform for further modification to tailor and synthesize complex nanostructured materials, for example, a nanotube-nanoparticle hybrid (Huang et al., 2004). Moreover, a practical pathway was thus provided to introduce specific chemical/biological properties into natural substrates. For instance, protein molecules were readily immobilized on the fine cellulose fibers of filter paper with the intermediation of the titania ultrathin layers, resulting in a cellulose derived biosensor material (Huang et al., 2006). More interestingly, organic, porous, hierarchical bio-inspired materials with precisely reproduced nanostructure can be obtained by dissolution of the template substrates under mild conditions, after formation of organic thin layers (Gu, Huang, 2009). A bulk, porous, hierarchical cellulose derived titania/polyethylene hybrid sheet was obtained after dissolving away a template of cellulosic substance with sodium hydroxide/urea aqueous solution. With further acidic treatment, the metal oxide component can be removed and a pure polyethylene, hierarchical material obtained; while the hierarchical structure and nanoscopic details inherited from the original cellulosic sheet is maintained. Apart from them, cellulose can also perform as a high mechanical scaffold or provide sufficient carbon resources through pyrolysis.

4.2 Natural Cellulosic Substances

Cellulose is a linear polysaccharide of β-(1→4)-D-glucopyranose which is ubiquitously used in the formation of wood, cotton, and other plant fibers as an energy source, for building materials, and for clothing. As illustrated in Fig.4.1, there are hydroxyl groups placed at positions C2 and C3 (secondary, equatorial), as well as C6 (primary). The CH_2OH side group is arranged in a trans-gauche position relative to the O5-C5 and C4-C5 bonds. Through acetal functions between the equatorial OH group of C4 and the C1 carbon atom, β-D-glucopyranose molecules are covalently linked with each other and biogenetically form an extensive, linear-chain polymer with a large number of hydroxy groups. The molecular structure imparts cellulose with various characteristic properties such as hydrophilicity, chirality, degradability, and broad chemical variability initiated by the high donor reactivity of the OH groups. Meanwhile, extensive hydrogen bond networks are formed on the basis of the abundant hydroxyl groups on the molecular surface, resulting in a defined hierarchical order of supramolecular organization as well as in the partially crystalline fiber structures and morphologies. The elementary fibrils (length 1.5~3.5 nm) assemble into microfibrils (length 10~30 nm, also called nanofibers) and further bundle to form a hierarchical randomly cross-linked network of microfibrillar bands (length in the micrometer scale or even larger). This unique complex architecture endows cellulosic substances with high mechanical properties, as well as other biological functions and versatile applications. Meanwhile, the sufficient hydroxyl groups provide hydrophilicity, while the large number of hydrogen bonds prevents the cellulose from dissolving in common solvents easily. The pore structure formed in the cross-linked network is considerably important for the accessibility in chemical reactions and enzymatic degradation. The controlled variation of pore structures provides cellulose products a wide range of applications, from highly specialized membranes and carrier materials to consumer goods, such as nonwovens with excellent absorption properties. Such fascinating properties make natural cellulosic materials to be widely used in industries and our daily lives (Klemm et al., 2005). Especially for material researchers, natural cellulosic substances are ideal template substrates for designation and preparation of advanced materials. It is noteworthy that cellulose is chemically inert but it is possible to be modified with other chemicals due to the vast amount of hydroxyl groups. Thus, natural cellulosic substrates are suitable for metal oxide film deposition via the surface sol–gel process.

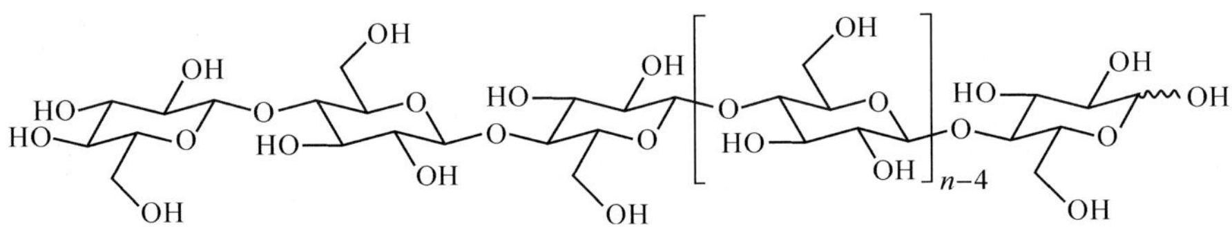

Fig.4.1. Molecular structure of cellulose (n=DP, degree of polymerization)

The surface sol–gel process is a facile and powerful chemical tool to form ultrathin metal oxide films with molecular precision. As schematically presented in Scheme 4.1a, a metal alkoxide precursor is first chemically adsorbed from the corresponding solution onto the hydroxyl terminated substrate surface to form a covalent bond, and then hydrolyzed to the metal oxide gels, which provide a new hydroxylated surface for successive further deposition. Since the characteristic film formation mechanism, an individual metal oxide layer can achieve subnanometer thickness through careful experimentally controlled conditions. By repeating the deposition-hydrolysis cycle, faithful replication of the original structures and morphologies at extremely precise levels with desired film thickness can be obtained (Huang et al., 2002a; 2002b; 2002c).

In this process, ultrathin metal oxide gel films are formed layer-by-layer coating on the morphologically complex surfaces of each cellulose nanofibers with nanoprecision. Thanks to the subnanometer thickness of the as-deposited films, the hierarchical and slight morphologies of the template substrates are retained in metal oxide films. Macroscopic as well as nanometer scale fossils can be obtained after removal of the organic substances through calcination. The resultant fossils appear to be hierarchical ceramic materials that possess the structures and morphologies faithfully inherited from the original templates. The ceramics are composed of randomly cross-linked nanotubes, as precise negative, hollow replicas of the initial cellulose fibers. Scheme 4.1b shows the TEM images of a virgin cellulose fiber of filter paper, a titania ultrathin film coated cellulose fiber, and a titania nanotube obtained after calcination of the titania coated fiber, respectively, demonstrating the real sample of the replication process.

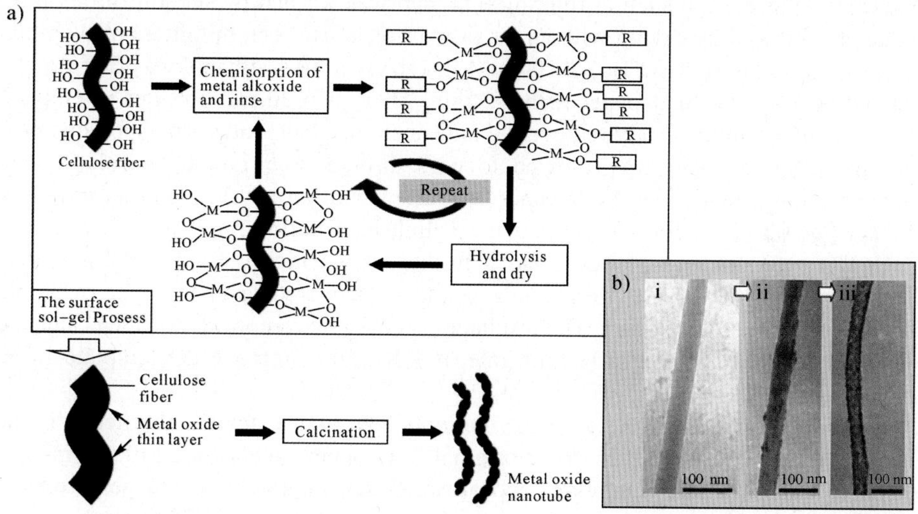

Scheme 4.1. a) Schematic illustration of the precise replication of natural cellulose fibers with metal oxides by the surface sol–gel process. Reprinted from Huang et al., 2005b. Copyright (2005), with permission from the American Chemical Society. b) Transmission electron micrograph (TEM) images: i. a virgin cellulose fiber of filter paper, ii. a titania-coated cellulose fiber, iii. a titania nanotube that was prepared by calcination of the titania-coated fiber; deposition of a titania thin film was repeated twenty times for this sample. Reprinted from Huang et al., 2005a. Copyright (2005), with permission from the Royal Society of Chemistry

4.3 Cellulose Derived Nanomaterials

To faithfully replicate the structure and morphology of the initial natural cellulosic substances, a series of guest materials such as titania, zirconia, tin oxide, and indium tin oxide, and polyethylene were employed and yielded various related functional materials.

4.3.1 Titania Nanotubular Materials

Natural filter paper, consisting of cellulose fibers only, was the template for titania nanotubular fossils that were fabricated through the deposition of titania gel films on the morphologically complex surface of cellulose fibers of the filter paper. This process was accompanied by the subsequent calcination to remove the original filter paper (Huang, Kunitake, 2003). In a typical procedure, a piece of commercial filter paper was placed in a suction filtering unit, and was washed by suction filtration of ethanol, followed by drying with an air flow prior to use. Ten milliliters of titanium *n*-butoxide solution ($Ti(O^{n}Bu)_4$), 100 mM in 1:1/v:v toluene/ethanol was first slowly passed through the filter paper within 2 min. Then, two 20-mL portions of ethanol were filtered to immediately remove the unreacted metal alkoxide. And subsequently, 20 mL of pure water was allowed to pass through to promote hydrolysis and condensation. Finally, the filter paper was dried with an air flow and an ultrathin titania gel film was formed coating the cellulose nanofiber, and provided a new hydroxyl terminated platform for further deposition. By repeating this filtration/deposition cycle, thin titania gel layers covered layer-by-layer over the surface of the cellulose fibers. The resultant paper/titania composite was then calcined in air at 723 K for 6 h at a heating rate of 1 K/min to remove the template filter paper.

As displayed in Fig.4.2a inset, the obtained titania fossil inherited all the morphological characteristics of the original filter paper except for a little shrinkage in size, this situation commonly appears after calcination of a sol–gel product. Similar to the filter paper, the resultant titania replica sheet is self-supporting and highly porous, while the thickness and mass were reduced to *ca.* 0.22 mm and *ca.* 1.5 mg, respectively. Obviously, the sheet size and thickness depend on the template of filter paper used.

As clearly observed in the SEM and TEM images shown in Figs.4.2a and 4.2b, the original morphology of the filter paper was faithfully copied by titania ultrathin films and the replica appeared as non-uniform isolated, as well as bundled nanotubes. While the outer diameter of the titania tube varies from 30 nm to 100 nm, the wall thickness is highly regular that about 10 nm. The wall thickness can be controlled by alternating the number of deposition cycles of titania layers due to the ultrathin nature of the individual films. The titania nanotube assembly manifests the original structure and morphology of inter woven cellulose fibers, and the nano-branched structure of the initial fibers is clearly visualized in Fig.4.2b. From the SAED pattern obtained from agglomerated titania tubes presented in the inset of Fig.4.2b, typical diffraction rings show the formation of an anatase crystal, revealing the titania nanotubes are composed of anatase fine particles with sizes of around 10 nm.

This replication process can be readily applied to other natural cellulosic substances such as cloth and cotton similarly, resulting in "titania cloth" and "titania cotton" as displayed in Figs.4.2c and 4.2d, respectively. The titania films again faithfully memorize the hierarchical morphologies and structures of the original bio-templates from macro-levels down to nanometer scales, and hence give macroscopic fossils. Fig.4.2c gives the fine titania thread copied from an individual fiber which makes up strands in the initial cloth, while Fig.4.2d presents the fine titania hair inherited from the spiral twist of natural cotton lint. They are composed of arrays of tortuous titania nanotubes as precise replications of cellulose fiber assemblies, both of which are clearly demonstrated in the corresponding high magnification SEM images (insets of Figs.4.2c and 4.2d). Although the initial paper, cloth, and cotton are all natural cellulosic substances, there exist some structural differences and thus the corresponding titania fossils appear to have topographic differences as displayed in Figs.4.2a, 4.2c and 4.2d.

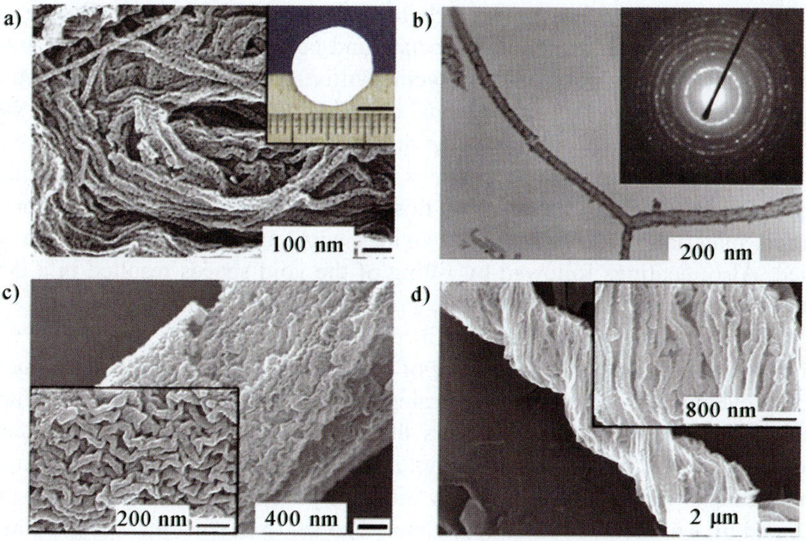

Fig.4.2. Titania fossils of natural cellulosic substances. Deposition cycle of titania thin films was repeated 20 times for each sample. a) Field emission scanning electron micrograph (FE-SEM) of "titania paper", showing titania nanotube assemblies. The inset shows photograph of a sheet of "titania paper". b) TEM image of individual titania nanotube isolated from the assembly; inset of b), selected-area electron diffraction (SAED) pattern from the nanotube assembly. c) FE-SEM image of "titania cloth". d) FE-SEM image of "titania cotton". Reprinted from Huang, Kunitake, 2003. Copyright (2003), with permission from the American Chemical Society

Among various reported oxidic nanotubes, the titania nanotube is especially fascinating because of its unique electronic, photonic, and catalytic properties other than low-cost, high stability, and non-toxicity. Currently, a series of ingenious and practical environment-friendly approaches are presented to design and fabricate nanotubes with a proper selection of template substrates. These features are not readily achieved by the traditional chemical methods such as sol–gel template syntheses employing porous membranes (Lakshmi et al., 1997; Liu et al., 2002) and polymer fibers (Caruso et al., 2001), or alkali treatment on titania powders (Kasuga et al., 1998; Chen et al., 2002).

It is worthy of mentioning that the titania nanotubes possess a multi-helical morphology which is recognized as important and a challenging unique class of advanced functional materials. As clearly displayed in Fig.4.2d, replication of natural helical architectures with inorganic matrices can provide an effective pathway for preparation of helical inorganic materials. It is known that every natural cotton hair is a thin, flattened, and tubular cell with a pronounced spiral twist when it is fully mature and dry; and its length reaches several centimeters. Faithful duplication of natural cotton hairs with titania, by the current facile strategy, endows "titania cotton" with multi-helical nanotube structures.

Attractively, the surface sol–gel chemistry is not limited to cellulose only. Cellulose derivates such as cellulose acetate and cellulose nitrate, as well as other organic membranes like polyamide, polyethersulfone, and polypropylene, were used as template substrates to fabricate porous, metal oxide films (Schattka et al., 2006). Since the initial substances remained, the original good mechanical properties can be kept as well. Moreover, multiple coatings can be readily applied with variation in the type of metal oxide precursor. With this "sequential coating" approach, complex structured materials of layered metal oxides such as titania and zirconia were obtained. Also, coatings followed by filling of the void spaces resulted in a unique titania coated, bimodal, and pored silica.

Since there exists surface tension in every liquid solution, penetration of the precursor solution is limited, and wetting of the smallest pockets of the nanostructures may be prevented. As an alternative strategy, gas-phase is used instead (Kemell et al., 2005). In ALD, film growth occurs through alternating saturations of surface reactions. Typically, the precursor vapors are pulsed into the reactor one at a time, and the precursor pulses are separated by inert gas purges. Then, the substrate surface is saturated by a monolayer of that precursor. Thanks to the fact that the gas-phase is free of the precursor, the next precursor selectively reacts with the adsorbed surface layer. Hence, the film growth occurs (sub)monolayer by (sub)monolayer, and the film thickness can be accurately controlled by changing the number of the deposition cycles. Because this growth mechanism is self-limited, the as-grown thin films are conformal, and their thicknesses and compositions are uniform over large areas. These amazing features make ALD a highly suitable method for conformal coating of natural substances that possess sophisticated structures. It is worthy of note that the temperature ranges for ALD growth are wide, and thus this strategy is applicable at low temperatures for natural fibers. For atomic layer deposition of titania ultrathin films, halides and alkoxides are chosen as precursors, and finally air-annealed to remove the natural cellulosic substance. Either the amorphous or crystalline phase of the resulting titania films depends on the deposition temperature. The resulted titania replica successfully memorized even the smallest details, such as a triangle-shaped intersection of three nanotubes, of the initial filter paper, and exhibited good photocatalytic activity. The wall thickness is about 30 nm. Considering 1,000 ALD cycles were repeated to form the titania film, the growth rate is only about 0.03 nm per cycle. If less total cycles can be realized, more precise replication will be possible.

4.3.2 Zirconia Nanotubular Materials

Since carbon nanotubes were discovered, great attentions have been paid to nanotubular materials in both fundamental and industrial researches due to their particularly distinguished properties like high surface-to-volume area and low density, which bring superior efficiencies and performances to the corresponding bulk materials and isotropic nanoparticles (Patzke et al., 2002; Xia et al., 2003). Unfortunately, traditional efficient synthetic approaches are not suitable for fabrication of oxidic nanotubes. As described above, the current "artificial fossilization" strategy provides a facile and powerful shortcut to prepare metal oxide nanotubular materials. Other than titania, metal oxides like zirconia are also employed as replicating matrices in the surface sol–gel process. In this case, zirconium *n*-butoxide was chosen as the precursor and a man-made natural cellulose paper derived zirconia nanotube fossil was successfully synthesized. The obtained product exhibited well aligned nanotubular structures precisely replicated from the initial cellulosic substance with a uniformly defined ultrathin wall thickness of about 10 nm and an extremely high aspect ratio (Huang, Kunitake, 2003).

4.3.3 Tin Oxide Nanotubular Materials

Tin oxide, a stable wide-band gap *n*-type semiconductor, is widely known as a promising key functional material for various practical applications, particularly for gas sensing. The nanotubular materials using this amazing metal oxide were also fabricated with a natural cellulosic substance (filter paper) as a template (Huang et al., 2005). $Sn(O^iPr)_4$ was chosen as a precursor compound to coat cellulose fibers of filter paper with SnO_2 gel layers by the surface sol–gel process first. Calcination in air followed, giving hollow SnO_2 nanotubular materials as natural cellulose fibers replicas. By repeating the surface sol–gel process deposition cycle for twelve times, and subsequently calcined at 723 K, the as-deposited amorphous-like sample, composed of fine particles with sizes smaller than *ca.* 5 nm, was obtained. The final product was a macroscopic fossil of the template filter paper and appeared as a self-supporting ceramic sheet. From the SEM and TEM images exhibited in Figs.4.3a~4.3d, the outer diameters can be observed as tens to two hundred nanometers and wall thickness are measured at 10~15 nm. If calcined at 1,373 K, tube-like polycrystalline SnO_2 nanocages composed of rutile-phase SnO_2 crystalline nanoparticles with 10~20 nm thickness are yielded, where the outer diameter is between 100~200 nm. On the other hand, Cotton fiber templated SnO_2 microtubes are prepared by a chemical deposition technique through oblation and heterogeneous nucleation of tin difluoride (Imai et al., 2000). Although in both cases tubular structures were achieved, the latter one was replicated only at the micrometer scale and the diameters and wall thicknesses of the resultant tubes were in the micrometer regime. While the bio-inspired SnO_2 tubules, fabricated through the surface sol–gel process, replicated the natural cellulose template at all levels of morphological hierarchies from nanometer to centimeter regimes.

The formed structures of SnO_2 nanotubes were strongly affected by the conditions of thermal treatment. As the XRD patterns of as-prepared SnO_2 powders calcined at different temperatures from 573 K to 1,173 K, exhibited in Fig.4.3e, though all diffraction peaks well meet the tetragonal structure (rutile type) in JCPDS 411445 (cassiterite) and indicate that the specimens consist of SnO_2. The intensity of each peak gradually increased with the increase of calcination temperatures. Especially, the sample obtained under calcination conditions above 773 K showed typical diffraction peaks of cassiterite, suggesting that the sample was composed of pure SnO_2 only without any other component. Besides, the crystallite size increased along with the calcination temperature gradually up to 973 K, and turned faster after this point. When the sample powder was calcined at 573 K, the resulted crystallite size was only about 2.0 nm, which is smaller than that of SnO_2 derived from $SnCl_4$ (about 4 nm) calcined under same conditions (Xu et al., 1991). Even when the calcination temperature reached 1,173 K, the crystallite size did not grow lager than 10 nm. TEM observation showed that sized of SnO_2 nanoparticles are about 5 nm at calcination temperature of 723 K and 10~20 nm at 1,373 K, respectively. These results are in fair agreement with nanoparticle sizes estimated for each calcination temperature from the XRD measurements, though the two sets of data for 1,373 K cannot be directly compared. In any case, the crystal growth is clearly suppressed even at high calcination temperatures. The mechanism of such suppressed crystal

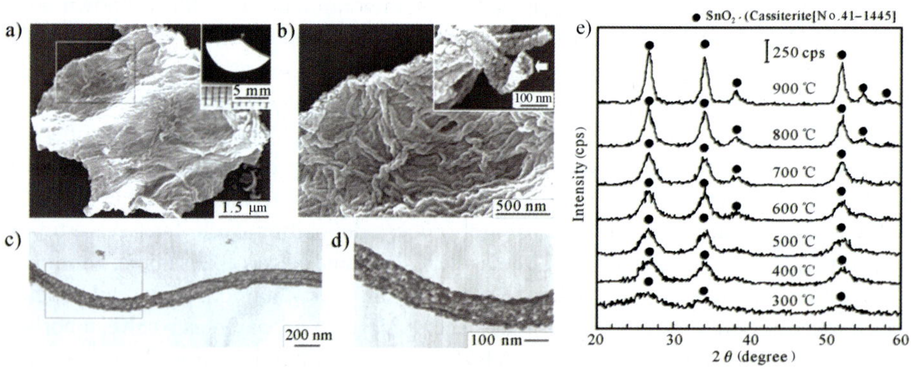

Fig.4.3. SnO_2 nanotubes prepared by calcination of a SnO_2 gel/filter paper composite sheet at 723 K. a) Low-magnification FE-SEM image of the SnO_2 nanotube assemblies. The inset is a macroscopic photograph of the sheet, which was yielded by calcination of a one-third of an as-prepared sheet. b) FE-SEM image of the morphological details of the boxed area in a). The inset shows high-magnification FE-SEM image of two individual SnO_2 nanotubes, the arrow indicates an opening of the nanotube. c) TEM image of one individual SnO_2 nanotube isolated from the assembly. d) High-magnification TEM image of the boxed area in c). e) X-ray diffraction (XRD) patterns of the nanotubular SnO_2 material obtained by calcination of the as-prepared SnO_2 sheet for 3 h at various temperatures. Reprinted from Huang et al., 2005. Copyright (2005), with permission from the American Chemical Society

growth of SnO_2 reported here is not clear, yet it is supposed that it may arise from the unique fabrication process of the SnO_2 sheet. It is considered that SnO_2 is formed separately on the cellulose fiber matrix and may possess limited contacts with the neighboring particles in the sheet.

The resulting SnO_2 nanotubes were endowed with several advantages due to the unique morphology and raw material, and hence showed great potential as novel gas sensors. While a thin-film type of SnO_2 sensor has a serious problem that the concentration gradients of sensing gases arises between the outermost layer of the film and the bottom layer directly attached to the electrode substrate. However, the high surface-to-volume ratio and the fine grain size of the obtained SnO_2 nanotubes increase the chance for the material to access the target gas molecules. The open nanotubular structure facilitates fast and full gas interaction to SnO_2 nanocrystals, resulting in the enhanced detection performance. Therefore, the tubule structured with SnO_2 is expected to possess better gas sensitivity and sensing reversibility compared with the traditional film or nanobelt type SnO_2 materials. Fig.4.4a displays corresponding response transients of a SnO_2 nanotube sheet sensor to 100×10^{-6} hydrogen gas, 100×10^{-6} carbon monoxide gas, and 20×10^{-6} ethylene oxide in the working temperature range of 623~773 K, respectively. The resistance (R_a) of the sensor in dry air was (1~2) $\times$ $10^6\,\Omega$ at the tested temperature and increased slightly with decreasing temperatures, which is close to that of SnO_2 crystallite (Baik et al., 2000). Among the sample gases, although a low concentration range was chosen for ethylene oxide, the response transient to the smallest hydrogen gas was the largest. For the sensitivity to target gas mainly depends on the ease of diffusion of gas molecules inside the sensor. The hydrogen molecule can diffuse most easily inside to the deeper regions of the sensor, to access the oxygen adsorbed on SnO_2 surface, resulting in the largest response. From Fig.4.4b, dependence of sensitivity (S) of the test gases versus the working temperatures can be observed. The sensitivity curve of hydrogen gas gives a concave shape with a peak appearing at 723 K, where S equals 16.5 and is comparable to the conventional SnO_2 sensor; while the S curves of the other two gases went downward along with temperature increases. It is a pity that these S values at 723 K were not so high when compared with other reported results (S values measured 500×10^{-6} to 800×10^{-6} H_2 and 1,150 to 800×10^{-6} CO at 623 K, respectively) (Baik et al., 2000). Since it is well known that the gas sensitivity is strongly affected by the crystallite size, especially when the grain size is smaller than 6 nm, the gas sensitivity of a SnO_2 gas sensor drastically increases (Xu et al., 1991). Thus, the gas sensitivity of the current SnO_2 is predicted to be possibly much higher than the observed value. For the present SnO_2 sheet sensor is composed of highly entangled nanotubes, the access of gas molecules to these SnO_2 nanotubes is assumed to be not so favored as it is toward isolated nanotubes and bundles of nanotubes. To secure efficient access of gaseous reactants, whether nanotubes have porous wall or open tube ends is considered essential. Unfortunately, in the present case, such favorable situations are apparently not achieved, and hence the sensitivity is not as good as that of the thin-film SnO_2 sensors. It is assumed that isolated and/or better aligned nanotube structures are needed for the current cellulose fiber derived SnO_2 nanotubular materials to get enhanced gas access.

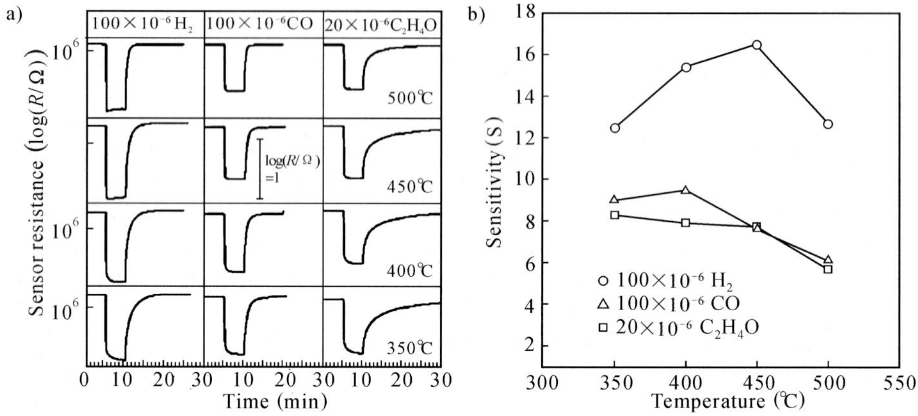

Fig.4.4. a) Response transients of SnO_2 nanotubes respectively to 100×10^{-6} H_2, 100×10^{-6} CO and 20×10^{-6} C_2H_4O at varied temperatures. b) Temperature dependence of sensitivities of SnO_2 nanotube sensor to 100×10^{-6} H_2, 100×10^{-6} CO and 20×10^{-6} C_2H_4O. Reprinted from Huang et al., 2005b. Copyright (2005), with permission from the American Chemical Society

4.3.4 Indium Tin Oxide Nanotubular Materials

Tin-doped indium oxide, called ITO, is a promising degenerate wide-band gap, *n*-type semiconductor which exhibits several characteristics such as relatively low resistivity, high optical transmittance in both the visible and near infrared regions, and high reflectance in the infrared region. It is these attractive properties that have made ITO to become the best-known transparent conductive oxide (TCO) material in optoelectronics. Then, what happens if ITO is made into nanotubular materials? Since it can possess nanotube topography and electronic conductivity, nanotubular ITO materials are able to offer unique chances to develop into functional devices and sensors. Although traditional investigations rarely involve this aspect, free-standing nanotubular ITO sheets with different In/Sn ratios were successfully fabricated with commercial filter paper as a template in applying the artificial fossilization process. The resulting materials faithfully memorized the hierarchical structure originating from the morphology of the cellulosic sheet. From nanoscopic observation, the ITO nanotubes are found to be composed of interconnected layers of ITO nanocrystals, whose diameters are just a few nanometers. As visualized in the inset of Fig.4.5a, the resulted pale-yellow ITO sheet is self-supporting, with highly similar morphological characteristics from macroscopic to microscopic sizes compared with the initial filter paper (Figs.4.5a and 4.5b). The SEM image shown in Fig.4.5a clearly indicates the product is composed of randomly cross-linked, irregular nanotube networks inherited from the original cellulose fiber network. As well as aligned ITO nanotube arrays led by the cellulose fiber assembly, the individual nanotubes can be clearly identified. The nanotubes own quite high aspect ratios, with outer diameters of a few tens of nanometers to about two hundred nanometers. Fig.4.5b presents the high-resolution FE-SEM images of the isolated ITO nanotube, demonstrating the tube is composed of nanoparticles whose sizes are only about 10 nm. The SAED analysis was carried out, which indicated the polycrystalline nature of the ITO nanotubes.

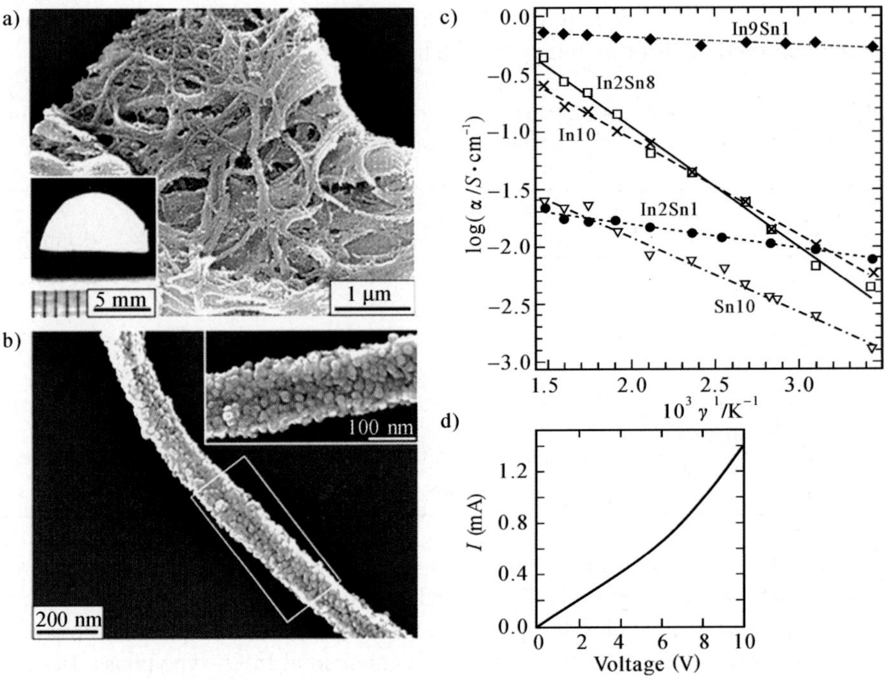

Fig.4.5. ITO (In2Sn1) nanotubes obtained by the artificial fossil process with commercial filter paper as template, deposition of ITO thin films was repeated 12 times for this sample. a) Low-magnification FE-SEM micrograph of the ITO nanotube sheet, showing nanotube networks; inset is a macroscopic photograph of the sheet, which was obtained by calcination of a half of an as-deposited ITO gel/filter paper composite sheet. b) FE-SEM image of one individual ITO nanotube isolated from the assembly, and the inset shows a high-magnification image of the boxed area. c) Arrhenius plots of the electrical conductivity of nanotubular ITO sheet. d) I-V curves of In9Sn1 at 293 K. Reprinted from Aoki et al., 2006. Copyright (2006), with permission from the Royal Society of Chemistry

ITO gel layers were deposited on cellulose nanofibers individually, employing indium methoxyethoxide and tetraisopropoxytin, at a total concentration of 12 mM, as a precursor solution. The effects of different In/Sn molar ratios were also investigated. The precursor mixtures of In/Sn ratios of 10/0, 9/1, 2/1, 2/8 and 0/10 were selected and respectively named hereafter as In10, In9Sn1, In2Sn1, In2Sn8, and Sn10 from the In/Sn ratios in the precursor solutions. To determine the practical In/Sn ratio in the nanotube sheets, electron probe micro analysis (EPMA) was made and the results are listed in Table 4.1. The observed In/Sn ratio is always a bit smaller than that of the precursor solution, which is probably caused by the relatively greater reactivity of the indium alkoxide compared with that of the tin alkoxide. By optical microscopy, the average thickness of the ITO sheet is measured in the range of 90~220 μm. On the other hand, the apparent density of the ITO sheet, determined from the nominal volume and weight of 10 mm × 10 mm pieces of the replica, appeared to be very low compared to that of the values located in the range of 1.8%~4.7% of the ideal density calculated from the bulk density of In_2O_3 and SnO_2 and the observed In/Sn ratio, indicating the highly porous structure of the fossils.

Table 4.1. Compositions and characteristics of ITO nanotubular sheets. Reprinted from Aoki et al., 2006. Copyright (2006), with permission from the Royal Society of Chemistry

Precursor mixture	In/Sn ratio	Thickness (μm)	Density ($g \cdot cm^{-3}$)	Fractional density (%)	σ_{298} ($S \cdot cm^{-1}$)	E_a (kJ·mol)
In10		98 ± 5	0.33	4.7	5.89×10^{-3}	16
In9Sn1	93.5/6.5 (14.4/1)	218 ± 12	0.23	3.3	0.533	1.2
In2Sn1	80.3/19.7 (4.1/1)	95 ± 11	0.13	1.9	7.58×10^{-3}	4.0
In2Sn8	30.4/69.6 (1/2.3)	172 ± 26	0.16	2.2	4.27×10^{-3}	20
Sn10		165 ± 19	0.13	1.8	1.03×10^{-3}	13

The XRD study towards the resultant powder product infers that the crystal phases of the ITO sheets varied with the different metal proportions. While the In10 sample consisted of the cubic In_2O_3 phase and a small amount of the rhombohedral In_2O_3 phase, the In9Sn1 is endowed with a rhombohedral In_2O_3-type phase. Because the ITO nanoparticles in the current sample consists of less than 10 nm in diameter, the preferential formation of a rhombohedral phase is considered to be due to the small particle-size. Meanwhile, both the In2Sn8 and Sn10 samples exhibit a single pattern of the tetragonal SnO_2-type phase, however, the peak width of the former one is broader than that of Sn10. On the other hand, In2Sn1 is demonstrated to be a mixture of rhombohedral $In_4Sn_3O_{12}$ and rhombohedral In_2O_3-type phases, though the relative intensity of the peaks are not as strong as those of others. The nanoparticles in the resulted ITO nanotube are crystalline ones and there is no phase separation happening between the indium and tin oxide.

To measure the electrical conductivity parallel to the free-standing ITO sheet surface, the four-probing van der Pauw method was employed. The obtained Arrhenius lots of electrical conductivity, σ, of related samples indicate, as a result, that all the ITO sheet samples are semiconducting along with the temperature (Fig.4.5c). Among them, In9Sn1 demonstrates the highest σ (0.53 $S \cdot cm^{-1}$) at ambient temperature and the least activation energy (E_a) of 1.2 $kJ \cdot mol^{-1}$ (Table 4.1). In2Sn1 shows a similar E_a value to that of In9Sn1, but the corresponding σ is lower than that of In9Sn1 by a factor of 10^2 over the test temperature range. Compared to In9Sn1 and In2Sn1, the other three samples show apparent temperature dependence of σ with a large E_a. The σ values of In10 and In2Sn8 are in the order of 10^{-3} S·cm^{-1} at 293 K, but will increase to about 0.3~0.4 S·cm^{-1} once heated to 673 K. While the conventional ITO materials like dense films, single crystals, and sintered pellets exhibit a metallic behavior at 300~700 K, and their conductivities decrease with increasing temperatures, the current ITO sheets reveal the opposite temperature dependence, which is assumed to be due to the different scattering mechanism. Fig.4.5d presents the I-V curve of the corresponding sample, and the one of In9Sn1 at room temperature is not linear in nature but shows an upward deviation from Ohm's law at the voltages higher than 5 V. Meanwhile, other ITO sheets also show a similar feature in I-V curves, which is attributed to the effect of charge trapping at the boundary. The grain boundary of polycrystalline semiconductors contains a great amount of point defects that induce trapping of the carriers, which produces the depletion layer with a potential barrier at the grain boundary, blocking the carrier motion from one crystalline to another. Such grain boundary scattering must significantly affect the current samples, for they are composed of the particle of just a few nanometers that is commensurate with the mean-free path that about 3 nm wide in nanocrystalline ITO films. In addition, the related electrical transport property is mainly determined by the nature of the grain boundary. The other scattering mechanisms such as crystallinity, the effect of low crystallinity may work but are considered rather small in this case. The σ value of In9Sn1 obtained at 293 K, measured as 0.53 S·cm^{-1}, is lower than that of the commercial ITO (5 × 10^3 S·cm^{-1}) by a factor of 10^4. However, if conductivity was corrected for the apparent density, the effective conductivity of solid In9Sn1 was calculated to be 160 S·cm^{-1}, which is only an order of magnitude lower than that of the commercial ITO film. On the other hand, when effective conductivities are compared, the effective electrical conductivity of the current free-standing ITO sheet is relatively higher than those of other nanostructured ITO materials.

The resulting high conductivity of the current ITO nanotube is assumed to be endowed by the unique structure. ITO gel layer is uniformly formed on the template of cellulose fibers, with nanometer precision, and is subsequently converted into nanotubes composed of interconnected nanocrystals through calcination. Even after calcination, the macroscopic morphology and the free-standing nature of the initial template filter paper are memorized, and hence the interconnection is extended from nanometer to macroscopic scales. This result infers that all the nanocrystals are covalently connected up to the macroscopic scale, as shown in Figs.4.6a and 4.6b, despite extremely small space occupation of the sample. Compared with the pelletized specimen of mesoporous ITO powder, where the electrical contact between the particles is achieved by physical contacts, the current material possesses the aforementioned structural property that facilitates the electron percolation. The covalent connection of nanocrystals may also make a contribution to reduce the grain boundary scattering, for the chemical bonding at grain boundaries can decrease the number of charge trapping sites like dangling bonds. The ITO sheet with In/Sn ratio of 93.5/6.5 demonstrated the highest effective electrical conductivity of 160 S·cm^{-1}, which is higher than those of the nanostructured ITO fabricated with other template synthesis process, as well as than that of the single-crystalline nanowhisker synthesized via vapor-liquid-solid (VLS) technique. The morphological and electrical characteristics in the resulted specimen originate from the covalent interconnection of ITO nanocrystals from the nanoscopic level up to the micrometer scale. The morphological and structural adaptability of the current ITO system is more distinguished than those of other conductive transition-metal oxides such as CuO, NiO, and MnO_2 which hardly give flexible nanoprecision morphologies, even though proper template structures are employed. Thus, the nanostructured ITO, like the single ITO nanotube, has great potential to be applied in nano/micro-sized electronic devices. From the macroscopic scale, the nanotubular ITO sheet, for instance, combines high electrical conductivity with high surface area and should have advantages as electrodes in various applications such as an electrochemical battery, electrochemical capacitor, electrolytic wastewater treatment, and in light electricity conversion systems. It is expected that the nanostructured ITO sheet can make significant contributions to the design and fabrication of electronic micro-devices and electrode materials with certain novel strategies.

4.3.5 Hybrid of Titania Nanotube and Gold Nanoparticle

The developed surface sol–gel process can not only be used to fabricate artificial fossils, but also "activate" the natural substances through metal oxide modification; and provide a natural scaffold and platform for the fabrication of functional composite nanostructured materials. Huang et al. gave an example of nanoparticle/nanotube hybrid material, in which certain guest nanoparticles are attached onto the wall of host nanotubes (Huang et al., 2004). For combining the unique structural features of nanotubes and the significant properties of nanoparticles, nanoparticle/nanotube hybrid materials are recognized to promise wide application ranges such as heterogeneous catalysis and molecular sensors. The titania nanotube and gold nanoparticle hybrid nanomaterials exemplified herein are composed of gold nanoparticles and titania nanotubes with the morphological hierarchy originated from the template of a cellulosic substance (filter paper). The inset of Fig.4.6a presents the resulted bulk material fabricated on the basis of the surface sol–gel process. Typically, 15 layers of negatively charged titania films were first deposited on the cellulose nanofibers as mentioned above, followed by adsorption of a monolayer of positively charged gold, and subsequently 5 more layers of titania film were additionally deposited coating the gold nanoparticles as well as titania covered cellulose fibers. Then, the as-deposited sample was subjected to calcination to remove the initial filter paper and the organic ligand on the gold nanoparticles, resulting in a self-upporting gold/titania hybrid sheet of a dark-brown color. This gold/titania composite sheet weighs about 2.4 mg and contains about 40% Au by weight.

The FE-SEM image given in Fig.4.6a indicates that the gold/titania hybrid sheet is composed of hierarchically and randomly cross-linked titania nanotubes whose surface is decorated with a large number of gold nanoparticles. The individual hybrid nanotube possesses a tube wall thickness of about 10 nm, and is uniform with an extremely high aspect ratio. And Fig.4.6c clearly shows that the gold nanoparticles are individually and uniformly attached onto the titania nanotube as a monolayer, and each gold nanoparticle is covered by an ultrathin titania film. The TEM images visualize a tubular structure as well as high coverage of the gold nanoparticles on the titania nanotube surface. The inter-particle distance is measured as *ca.* 3 nm, which is apparently determined by the presence of the long chain organic ligand on the gold nanoparticle employed.

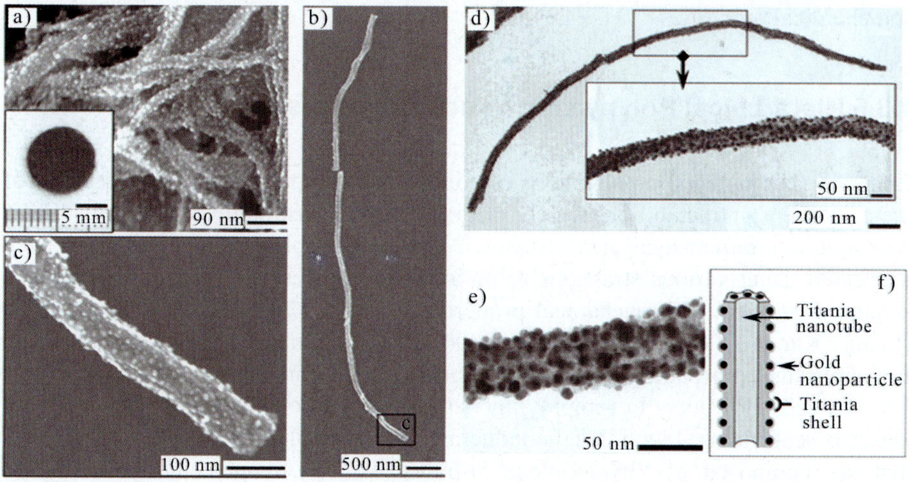

Fig.4.6. A hierarchical hybrid of gold nanoparticles and titania nanotubes, $[(TiO_2)_{15}$/Au-nanoparticle/$(TiO_2)_5]$, as derived from a cellulosic sheet. a) FE-SEM image of the hybrid, the inset shows a macroscopic photograph of the hybrid. b) FE-SEM image of an individual titania nanotube with gold nanoparticles anchored onto it. c) FE-SEM image of the details of the boxed area in b). d) TEM image of an isolated titania nanotube that is fully coated with gold nanoparticles, the inset shows the details of the boxed area. e) High magnification TEM image of the hybrid nanotube. f) Schematic illustration of the gold nanoparticle/titania nanotube hybrid (not to scale). Reprinted from Huang et al., 2004). Copyright (2004), with permission from the Royal Society of Chemistry

The additionally deposited titania thin layer on the gold nanoparticle makes them individually and wholly covered by the titania layer. It is known that gold nanoparticles undergo melting at relatively low temperatures (Ercolessi et al., 1991), which facilitates the fusion of the unprotected nanoparticles. The fusion was observed after heating for 30 s at 573 K in the case of gold nanoparticles (sizes, (6±1) nm) on a carbon nanotube (Fullam et al., 2000). The titania layers surround the individual gold particles and inhibits fusion with adjacent gold particles even at higher temperatures. In the case of this sample $[(TiO_2)_{15}$/Au nanoparticle/$(TiO_2)_5]$, the gold nanoparticles are protected by 5 titania layers' coating (thickness about 2.5 nm), the average size and the standard deviation of the coated particle are 4.9 nm and 1.4 nm, respectively. Actually, the original size distribution, (5±1) nm, is not changed even after a long period of calcination (6 h at 723 K). On the contrary, the nanoparticle without titania layer protection was observed to undergo particle fusion. For the hierarchical hybrid material of randomly cross-linked gold nanoparticle covered titania nanotubes, enhanced particle stability in the hierarchical morphology can be guaranteed, besides high and uniform metal loading and large surface areas. The one-pot fabrication of such sophisticated loading matrices is rendered applicable by the proper design of hierarchical templates, and should be strongly beneficial from the practical standpoint. The present approach, combining the rich varieties of nanoparticles and ceramic nanotubes, can produce various nano-precise systems with unique physical and chemical properties.

4.3.6 Hierarchical Polypyrrole Nanocomposites

Generally, the potential applications of conducting polymers are limited because of their inherent intractability, which originates from insolubility and infusibility. Although one-dimensional nanostructured conducting polymer materials have been fabricated using various strategies to optimize the process's ability and to achieve enhanced physical and mechanical properties (Martin, 1994; Carswell et al., 2003; Huang, Kaner, 2004; Zhang et al., 2004), selective micro- to nanometer scale deposition and morphology control of nanostructured conducting polymers are still not realized. However, by employing natural cellulosic substances such as filter paper as scaffolding, hierarchical conducting polymer polypyrrole (PPy) composites that are composed of PPy/cellulose bi-hybrid or PPy/titania/cellulose tri-hybrid nanocables were successfully obtained (Huang et al., 2005a). This result provides a novel pathway to design and fabricate artificial polymer nanomaterials.

To fabricate the target conducting polymer materials, ultrathin PPy layers were deposited onto each cellulose micro- and nanofibers of a piece of templated filter paper through an *in situ* oxidative polymerization process by immersing the filter paper into the prepared extremely dilute pyrrole solution (Huang et al., 2005a). The inset of Fig.4.7a is a photograph of the as-prepared PPy-cellulose composite paper sheet, indicating the macromorphology and flexibility of the initial filter paper is maintained. The SEM image shown in Fig.4.7a clearly demonstrates the continuous randomly interconnected network of nanofiber assemblies of the sheet, which are faithfully inherited from the template substrate. When the nanofiber is heated by focusing the electron beam used in the SEM system on it, it is observed to expand remarkably along the long axis, sometimes even to the point where the PPy layer is ruptured as displayed in Fig.4.7b. From the ruptured point of the PPy layer, a core-shell structure can be clearly seen. The bi-hybrid nanostructure composed of a cellulose core and PPy sheath is not a partial one, on the contrary, each fiber appeared as such a nanocable as the TEM micrograph presented in Fig.4.7b. An amorphous PPy coating ultrathin layer is seamlessly adsorbed onto the cellulose fiber, and grows into the PPy film which is uniform and homogeneous with a flat surface. The PPy sheath thickness measured about 20 nm, and the thickness of the PPy film can be precisely controlled by changing the adsorption and oxidization times. The resulting hierarchical PPy-based composite sheet possesses several useful properties such as high mechanical strength and a large surface area of PPy, which is possible thanks to the existence of the cellulose network and the porous morphology, respectively.

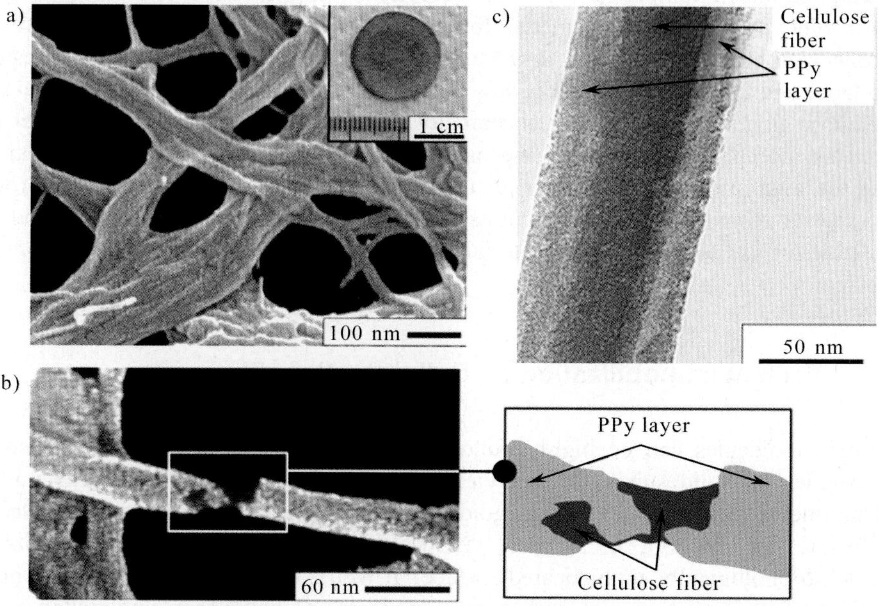

Fig.4.7. PPy-coated filter paper. a) FE-SEM image of the sample, visualizing the fibrous assembly; the inset is a photograph of the bulk sheet. b) FE-SEM image of a PPy-coated cellulose fiber which was broken by extended exposure to the electron beam; morphology of the broken part is schematically illustrated in the right inset. c) TEM image of a PPy-coated cellulose fiber. Reprinted from Huang et al., 2005a. Copyright (2005), with permission from the Royal Society of Chemistry

Cellulose has been expected to add strengthening fibers for new polymeric composite materials. To well accomplish this purpose, modifications of cellulose surface properties are generally necessary (Carlmark, Malmström, 2002). However, it is difficult because of the complex structures and relatively inert cellulose fiber surface. The present natural cellulose fiber derived composite paper sheets can successfully overcome these limitations and exhibit several interesting surface behaviors apart from the unique hierarchical morphologies and the nanocable structures. Neat filter paper is known to be extremely hydrophilic due to its absorbing nature, and the PPy/cellulose paper sheet inherits this behavior. On the other hand, a titania/cellulose paper sheet shows remarkably increased hydrophobicity, so that the contact angle of water on it is found to be about 121°. When PPy layer is further coated on the titania modified cellulose fibers, the resulting hybrid paper sheet exhibits less hydrophobicity, the corresponding contact angle is reduced to about 51°. Therefore, the current nano-coating approach opens a facile pathway to tailor cellulose fiber surface properties, which is considered an important potential in fabricating plastic composites.

Instead of PPy, Mariano et al. (2005) reported a porous carbon-carbon composite through pyrolysis of resorcinol-formaldehyde (RF) resins formed on a natural cellulosic template. The resins obtained by condensation of resorcinol with formaldehyde, in an aqueous media, form sub-micrometer sized clusters and further produce a highly porous gel (Al-Muhtaseb, Ritter, 2003). Usually such resins collapse and the porosity is lost if they are simply dried in the air. However, natural cellulose fibers are able to induce porosity by templating or stabilizing the gel as well as affecting the pyrolysis mechanism; in that way high porosity and large surface area are maintained through the air-drying process. The obtained porous composite resin, dried in air, exhibits differentiated ion exchange activity and is suitable for use as an electrode material in super-capacitors and other energy storage devices.

4.3.7 Protein Immobilization on Cellulose Nanofibers

Protein molecules can be highly biologically active, and thus immobilizing such molecules on solid substrates can yield bioactive functional surfaces. Generally, two-dimensional flat plates such as gold, silicon, and glass are employed as the solid substrates for immobilization. Then, what would happen if a protein is immobilized on a morphologically sophisticated surface? It would afford vast superiority over a solid surfaced substrate. Recently, some explorations have been undertaken with nanostructured materials such as nanoparticles and nanowires/nanotubes as new substrates (Cui et al., 2001; Chen et al., 2003; Kam, Dai, 2005). These novel bio-nanomaterials exhibit a great potential for developing new diagnostic strategies, drug delivery systems, and biosensor technologies. However, since an effective immobilization technique was rarely established, hierarchical mesoscopic scaffolds, ranging from macroscopic down to nanometer scales, have not been attempted as substrates. With an extension of the "artificial fossilization process", this limitation has been broken through and protein immobilization was achieved on natural cellulosic substances (*e.g.* filter paper) as the three-dimensional matrix. Protein molecules were successfully anchored on a natural cellulose nanofiber surface, giving a new class of bioactive nanomaterials.

In the typical fabrication process, the cellulose fibers of the filter paper were first coated by the nanometer-thick titania gel film via the surface sol–gel process, providing a biocompatible surface. Then, it was converted into a biotinylated one for protein immobilization. Protein (streptavidin) molecules were thereafter anchored through the high-affinity, biospecific biotin-streptavidin interaction as shown in Fig.4.8a. Control experiments for the cellulose sheets without a titania coating revealed that the streptavidin was not anchored at all, indicating the titania thin layer plays a key role here. The titania ultrathin film coated onto the cellulose fiber

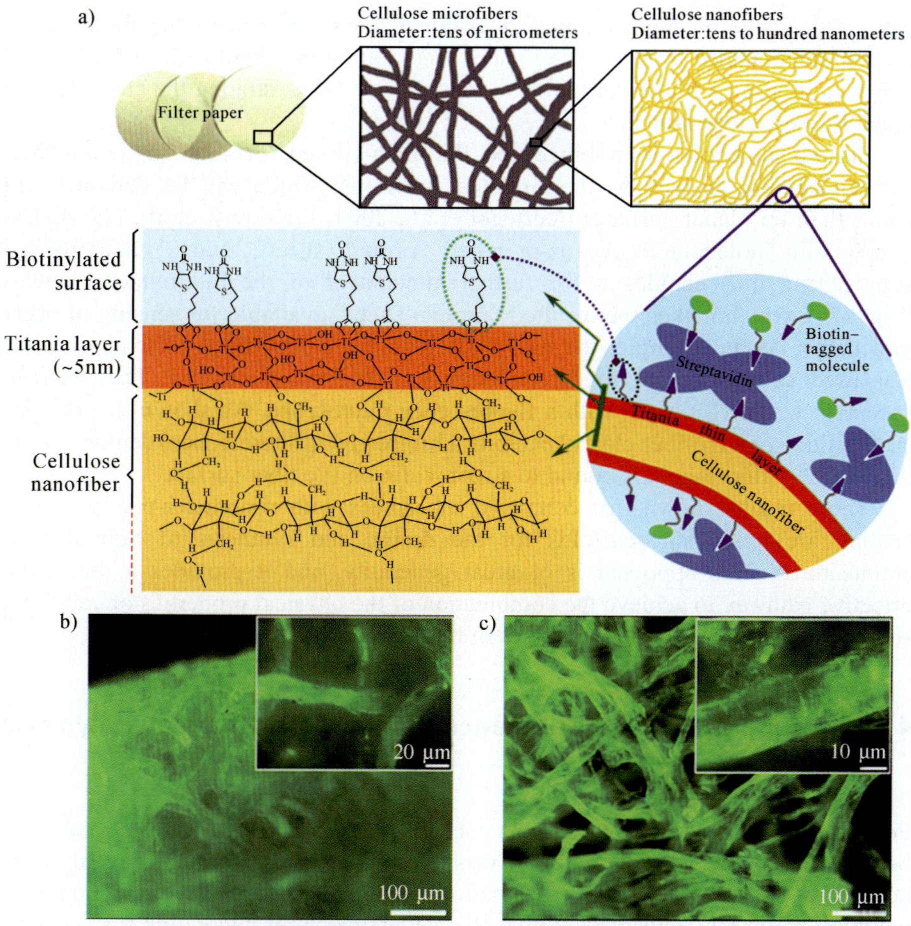

Fig.4.8. a) Schematic representation of immobilization of protein molecules (streptavidin) on each cellulose nanofibers and their successive binding to fluorescence-labeled biotin (not to scale). Left panel shows the formation of a biotinylated surface on a cellulose nanofiber; and right panel shows subsequent binding of streptavidin to the biotin-tagged species. The thickness of the titania layer is about 5 nm. b) Fluorescence micrographs of filter paper in which the cellulose fibers were decorated with Alexa 488-labeled streptavidin. c) Fluorescent micrograph of native streptavidin-modified cellulose fibers upon binding with biotin-4-fluorescein. Reprinted from Huang et al., 2006. Copyright (2006), with permission from Wiley

surface provides an activated surface for the formation of the biotin monolayer, and thus enables further immobilization of the streptavidin molecules. With this versatile approach, protein immobilization on solid surfaces is no more limited to two-dimensional flat substrates and nanostructured matrices with simple morphology. Due to the hierarchical porous structure and the high surface areas that are originated from the cellulosic matrix, the resulted protein sheet can act as a unique biomaterial for a wide range of applications. For example, Fig.4.8b presents the fluorescence micrograph obtained from the filter paper sheet immobilized with Alexa 488-labeled streptavidin using a fluorescence microscope, clearly demonstrating the characteristic green fluorescence emitted.

Protein molecules immobilized on the nano-curved surfaces can retain their native structure as well as their related functions, which can be demonstrated better than on planar surfaces (Vertegel et al., 2004; Lundqvist et al., 2004). The streptavidin molecule in the as-prepared "protein sheet" retains its biological activity and thus enables it for further attachment of the biotinylated species (Fig.4.8a), which makes the resulted "protein sheet" available for sensing of other molecules. As an example, the native streptavidin-immobilized filter paper sheet was used as a biosensor to detect biotin-4-fluorescein molecules. The resultant specimen was investigated with a fluorescence microscope. As shown in Fig.4.8c, each cellulose microfiber clearly exhibits a bright green fluorescence, demonstrating that biotin-4-fluorescein is bound to streptavidin on the fiber surface.

These results shown that cellulosic substances with nanometer-precise titania coating are promising scaffolds for the design and synthesis of biomolecular architectures. This approach is of great generality, and it provides a facile and effective pathway to achieve the combination of the physical properties of cellulosic substances and the high functionalities of bio-macromolecules.

4.3.8 Natural Cellulose Substance Derived Hierarchical Polymeric Materials

Porous nanostructured materials have been attracting general interests, and have become one of the hottest topics in versatile areas of research and development, because of their great potential as functional materials. Especially, the nanoporous polymer/hybrid materials present diverse open architectures and a high surface area. They further exhibit high charge/discharge capacity, high rate of adsorption, and efficient transport processes, both within and across the polymer walls. They also have the intrinsic properties of the raw organic materials which can realize potential applications in a rapidly increasing range of fields including electronics, optics, catalysis, separation, storage, sensors, and pharmacy (Xiao et al., 2007; Hassani et al., 2008; Welbes, Borovik, 2005; El-Zahab et al., 2004; Han et al., 2008; Lin et al., 2006; Tanaka et al., 2008; Morris, Wheatley, 2008; Shan et al., 2006; Ekanayake et al., 2007; Perez, Crooks, 2004; Li et al., 2003). It can be deduced that polymeric materials prepared by replication of natural substances are possibly endowed with interesting functions that originate from the unique template structures, which have a great application potential in medical care, cosmetics, electrochemistry, and various imaginable fields. Unfortunately, the template substances are generally removed by calcination, which severely limits the preparation of thermal unstable nanomaterials such as organics and polymers. Natural cellulose substances of templated polymeric/hybrid materials were realized by a layer-by-layer deposition of metal oxide/polymer thin films on cellulose nanofibers, with the successive dissolution of the initial cellulose template under mild conditions (Gu, Huang, 2009). The resultant natural cellulose derived hierarchical, titania/polymer, nanocomposite sheets exhibit high flexibility and mechanical strength, and show good absorption behavior similar to the initial cellulose templates. Furthermore, by subsequent acidic treatment to remove the titania component, these titania/polymer hybrid sheets can be converted into nanoporous pure polymer sheets.

The natural cellulosic sheet (filter paper) was employed as the template substrate. Firstly, ultrathin titania films were deposited using $Ti(O^nBu)_4$ as a precursor, providing a platform for polymer modification. Then, a poly(vinyl alcohol) (PVA) thin layer was subsequently deposited, forming a titania/PVA bilayer coating each cellulose nanofiber surface. Since there are abundant hydroxyl groups contained in PVA molecules as well as titania molecules, the raw materials are firmly linked with each other through covalent bonds and further deposition is possible. By repeating the titania/PVA deposition cycle n times, $(\text{titania/PVA})_n$ nanocomposite films were formed coating on each cellulose nanofiber surface of the filter paper.

Subsequently, to dissolve away the cellulose template without damaging other components, the resultant specimen was subjected to a sodium hydroxide/urea solution treatment under mild conditions. Under low temperature, the cellulose sheet can swell and promote the sodium hydroxide/urea solution to permeate and break inter- and intra-molecular hydrogen bonds, and form a large inclusion complex associated with cellulose, sodium hydroxide, urea, and water clusters. Finally, the cellulose was allowed to be dissolved (Cai, Zhang, 2005; Zhou et al., 2007; Cai et al., 2007). This method enables the removal of cellulosic substances under a moderate environment, and thus cellulose derived, porous, nanostructured products consisting of thermal unstable materials can be successfully obtained.

As control experiments, synthesis of a titania sheet and a titania/poly(acrylic acid) (PAA) hybrid sheet were also attempted by means of the same strategy. Unfortunately, the specimens were fragmented during the dissolution process. Compared with titania/PVA hybrid sheets, the mechanical property of the titania sheet is not strong enough and the PAA thin layers, which consisted of the titania/PAA hybrid sheet, were not stable in strong alkali solutions. Therefore, the dissolution process using a sodium hydroxide/urea solution is only considered to be useful when the required polymeteric materials possess a moderate degree of mechanical strength and are stable in an alkali environment; so that they can realize dissolution of the cellulose without any other deleterious influence on their structure.

After dissolving the cellulose template component, bulk titania/polymer hybrid sheets, as shown in the insets of Figs.4.9b1 and 4.9c1, were yielded. The resulting bulk, self-supporting $(\text{titania/PVA})_{10}$ sheet (the walls of the tubes consist of ten cycles of titania/PVA bilayer, similarly denoted hereafter) possessed the highly similar structural and morphological characteristics to that of the original cellulose sheet, except for a little shrinkage in size, which is considered to be due to the flexibility and elasticity endowed by PVA component. The SEM images displayed in Fig.4.9 directly indicate that the obtained hybrid sheets faithfully memorized the macro and microscopic structures as well as nano-precision details. Besides the randomly cross-linked hierarchical network of structures, thanks to the PVA, the resulted bulk sheets exhibited strong mechanical properties so that they were not easily destroyed by the grinding and extended sonication treatment. Meanwhile, a calcination process was avoided and as a result titania remained in an amorphous state, which also contributed to improved flexibility. Figs.4.9a2, 4.9b2, and 4.9c2 present the close-up views of the corresponding specimens, indicating that the microstructures of titania/PVA hybrid sheets are the same as that of the initial filter paper. The insets of Figs.4.9a2, 4.9b2, and 4.9c2 are TEM micrographs of individual tubes isolated from the corresponding specimens, demonstrating that the bulk

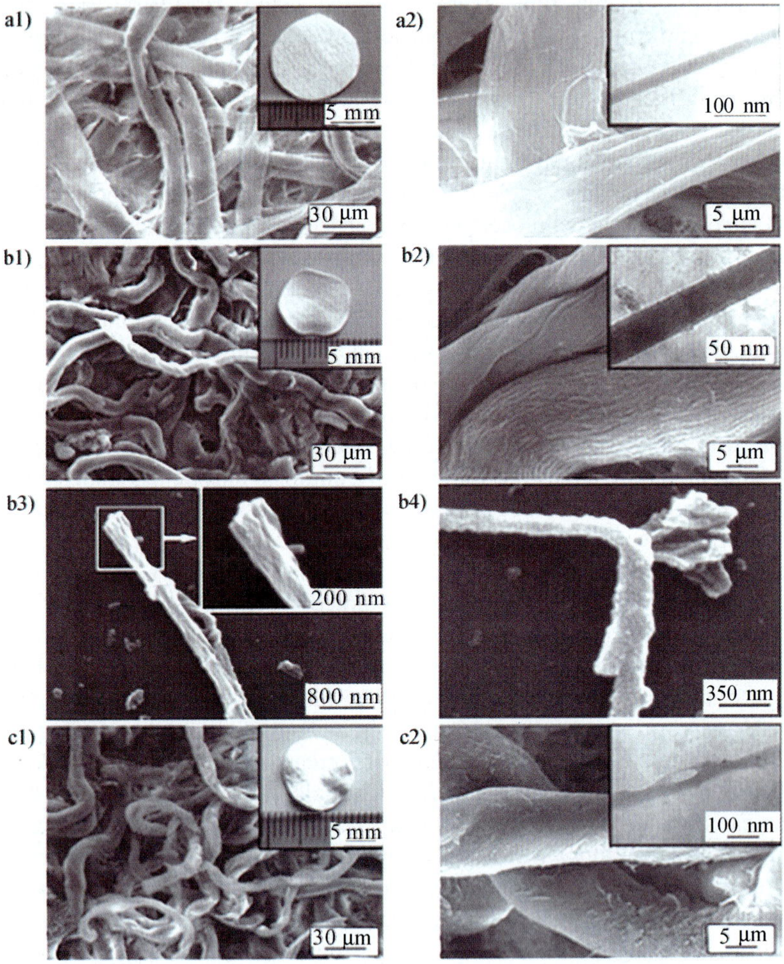

Fig.4.9. SEM micrographs of bare filter paper and hierarchical titania/PVA hybrid sheets. a1), a2) FE-SEM images showing the structure of bare filter paper. b1), b2) FE-SEM images of porous $(titania/PVA)_{10}$ sheet. b3), b4) FE-SEM images of $(titania/PVA)_{10}$ nanotubes. c1), c2) FE-SEM images of $(titania/PVA)_5$ sheet. And the insets of a1), b1), and c1) are the photographs of the corresponding dry bulk sheets. The insets of a2), b2), and c2) are TEM of individual nanotubes isolated from the corresponding specimens. The inset in b3) shows enlargement of the boxed area. Reprinted from Gu, Huang, 2009. Copyright (2009), with permission from the Royal Society of Chemistry

titania/PVA sheets obtained have nanotubular structures.Moreover, the nanotubes were flexible rather than just rod-type. It also infers that the tubing wall thickness is measured as *ca.* 30 nm, if Au-coatings to enable SEM observation are taken into consideration. Compared with that, nanotubes consisted of 20 pure titania layers whose wall thickness is *ca.* 10 nm (Huang, Kunitake, 2003), the $(titania/PVA)_{10}$ wall is much thicker, which is assumed to be due to the different self-assembly behaviors of PVA and titania.

These results indicate that the present products not only negatively replicated the original structure of the initial cellulose substance, but also possess a highly porous architecture and various functions originated from the polymer component. In addition, a bulk nanostructured, porous, PVA sheet can be obtained by treating the resultant titania/PVA nanocomposite sheet with an acidic solution (pH=1.0) to selectively remove the titania component. The resulted pure polymer product retains the overall hierarchical structure of the initial filter paper template as well as the original nanotubular morphologies. For the specimens with fewer polymer layers, it is observed that the mechanical property is relatively weak due to the thinner wall thickness.

Cellulosic substances are extensively utilized in industries as well as our daily lives not only due to their strong mechanical property and good flexibility but also because they swell, but do not dissolve in commonly used solvents, which makes cellulose known as an absorbent material. Interestingly, the currently fabricated titania/PVA, nanocomposite sheets memorized the absorbent behavior of the cellulose sheets. It is assumed that the titania/PVA specimens swell and absorb solvents following the same mechanism as that of bare filter paper, which mainly depends on physical absorption and penetration, hydrophilic properties, and sheet weight. For the hydrophilic properties of both PVA and titania, which were composed of the specimens, were not as good as those of celluloses, the titania/PVA sheets also exhibited a poorer swelling degree compared to that of the bare filter paper. It is worth mentioning that the swollen weight of a $(titania/PVA)_5$ composite sheet almost doubles that of $(titania/PVA)_{10}$ composite sheet. Considering the structure and hydrophilicity of the specimens are merely influenced by the bilayer number, the absorbed solvent volume should be quite close. Thus the weight gain resulting from increasing the layer number may be the main reason.

With this mild process, natural cellulosic substances for templated materials can be extended into organic materials. This methodology sheds a considerable light on tailoring and fabrication of bio-inspired organic, hierarchical, nanostructured materials.

4.3.9 Metal-coated Cellulose Fibers

The use of cellulose fibers as material scaffolds is also ideal because the fiber itself is tough and resilient, as well as being applicable to the electroless chemistry process. And such cellulose fiber supported conductive metal composites demonstrate various characteristics such as robust, resilient, lightweight, easy producing process, low-cost, and having uniquely significant dielectric properties with a low concentration of metal compared to the use of metal powders (Zabetakis et al., 2005).

The fabrication process of a cellulose-based dielectric composite is composed of four steps. Firstly, the cellulose fibers are fully hydrated to prevent excessive absorption of the chemical reagents. Secondly, a palladium compound is employed to activate the cellulose surface for metal plating. Subsequently, excess palladium and reagents used in activation were removed by extensive washing with water. Then, the product is freeze-dried to yield a fine gray powder, whose color is originated from the bound palladium. Finally, the cellulose fibers are metalized with an electroless copper plating solution, washed, and dried again.

The average longest axis for the length dimension of the resulting composites was 278 μm, and the original structures and morphologies of the cellulose fibers are well retained. Different from the imaginary component, the real component of the permittivity increases much faster with metal loading. Besides, the resultant composites demonstrate the unique dielectric properties with low-to-high dielectric constants, frequency dependence, and resonance effects. The high conductivity yet low density of the fibers allow the formulation of composites with a fraction of the weight of traditional microwave materials. As an alternative strategy, cellulose was used as a template substrate to support a metal-oxide catalyst (Shigapov et al., 2001). Several types of metal-oxide, including Ce-Zr mixed oxides and La-stabilized alumina, were formed onto the natural cellulose support, retaining a high-surface area, mesoporous structure with unusually high thermal stability. Thus, the application range is extended when catalyst meets natural supports.

4.3.10 Hierarchical Titanium Carbide from Titania-coated Cellulose Paper

Transition-metal carbides are focused as key materials with several practical uses in many areas, such as the cutting tool and abrasives industries, due to their hardness and chemical stability at high temperatures. Especially, titanium carbide (TiC) is one example of a high-temperature structural material with extreme hardness, low density, high thermal and electrical conductivity, and high mechanical stiffness. While the traditional methods to prepare such amazing materials have several deficiencies, such as relatively expensive starting materials, the products are frequently contaminated by a high oxygen content and particularly high temperature due to a high kinetic barrier. The titania ultrathin film coated cellulosic sheets, as described previously, provides a pathway to fabricate TiC materials under relatively low temperatures. With cellulose structures acting as a carbon precursor and an as-deposited titania component, TiC nanoparticles were synthesized by carbothermal reduction in Ar as in the following equitation (Shin et al., 2004).

$$TiO_2\,(s) + 3C\,(s) \rightarrow TiC\,(s) + 2CO\,(g) \quad (4.1)$$

Generally, this reaction requires higher temperatures because there exists a high kinetic barrier. However, with the unique ultrathin layer of titania deposited on cellulose nanofibers, the contact area between titania and carbon increased, and thus a considerably lower reaction temperature was realized.

SEM images of the titania-impregnated paper, calcined at 1,273 K in air, is presented in Fig.4.10a, showing that the resident cellulose structures are maintained during the formation of the rutile phase of titania. The TiC that formed at 1,773 K is a highly crystalline cubic form of TiC nanoparticles and the particle sizes are on the order of 10~50 nm, which are loosely agglomerated. The initial hierarchical structure of the attendant cellulose is faithfully replicated, which can be clearly seen in Figs.4.10b and 4.10c. While the products synthesized at a temperature lower than 1,473 K kept a high surface area, the surface area began to drop along an inverse correlation with the calcination temperatures increase above this point. This is assumed to be due to the increase of crystallinity and the collapse of the microporous structure, even under conditions where the hierarchical structures still remain. This idea lights up the extensive application of the natural template synthesis.

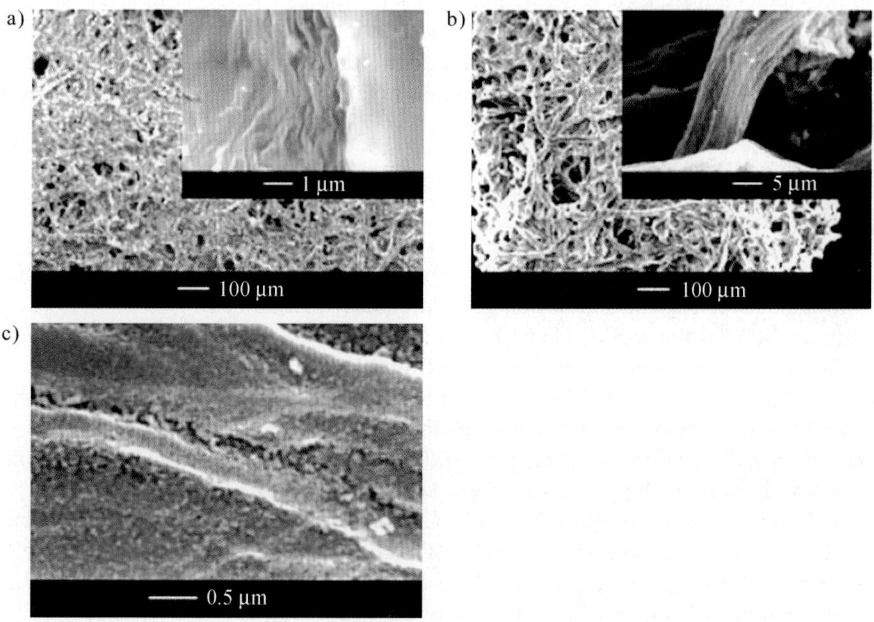

Fig.4.10. a) The SEM image of the titania paper calcined at 1,273 K in air. b), c) Highly crystalline TiC with replicated cellulose structures prepared at 1,773 K. The insets show high magnification images of the structures. Reprinted from Shin et al., 2004. Copyright (2004), with permission from Wiley

4.4 Summary

In conclusion, various advanced functional materials were successfully fabricated employing cellulosic substances as templates. A general chemical procedure called the surface sol–gel process was developed for the faithful replication from macro-scale down to nanoscale with metal oxides. By allowing hollow replication on both the micron and the nanometer levels simultaneously, this methodology widely extends the range of existing replication techniques and provides both positive and negative replicas of targeted objects with nanometer precision. It opens an effective pathway to prepare functional nanostructured products with intricate three dimensional morphologies. Also, it suggests a short-cut to probe structures of biosystems with extremely high precision. Therefore, the methodology introduced here is a practical, low-cost, and environmentally benign route to synthesize not only ceramic material but also organic products with unique tubular nanostructures. Moreover, this method can provide cellulose as a versatile performer for things such as templates, scaffolds and as carbon resources for further modification. New liquid phase processing methodologies such as a nanocopy technique (Kunitake, Fujikawa, 2003) and other strategies like the chemical vapor deposition (CVD) method (Kemell et al., 2005) are now in great demand.

References

Al-Muhtaseb SA, Ritter JA (2003) Preparation and properties of resorcinol-formaldehyde organic and carbon gels. Adv Mater 15:101-114

Anderson MW, Holmes SM, Hanif N, Cundy CS (2000) Hierarchical pore structures through diatom zeolitization. Angew Chem Int Ed 39:2707-2710

Aoki Y, Huang J, Kunitake T (2006) Electro-conductive nanotubular sheet of indium tin oxide as fabricated from the cellulose template. J Mater Chem 16:292-297

Baik NS, Sakai G, Miura N, Yamazoe N (2000) Hydrothermally treated sol solution of tin oxide for thin-film gas sensor. Sens Actuat B 63:74-79

Bao Z, Weatherspoon MR, Shian S, Cai Y, Graham PD, Allan SM, Ahmad G, Dickerson MB, Church BC, Kang Z, Abernathy HW III, Summers CJ, Liu M, Sandhage KH (2007) Chemical reduction of three-dimensional silica micro-assemblies into microporous silicon replicas. Nature 446:172-175

Cai J, Zhang L (2005) Rapid dissolution of cellulose in LiOH/urea and NaOH/urea aqueous solutions. Macromol Biosci 5:539-548

Cai J, Zhang L, Zhou J, Qi H, Chen H, Kondo T, Chen X, Chu B (2007) Multifilament fibers based on dissolution of cellulose in NaOH/urea aqueous solution: structure and properties. Adv Mater 19:821-825

Carlmark A, Malmström E (2002) Atom transfer radical polymerization from cellulose fibers at ambient temperature. J Am Chem Soc 124:900-901

Carswell ADW, O'Rear EA, Grady BP (2003) Adsorbed surfactants as templates for the synthesis of morphologically controlled polyaniline and polypyrrole nanostructures on flat surfaces: from spheres to wires to flat films. J Am Chem Soc 125: 14793-

14800

Caruso RA (2004) Micrometer-to-nanometer replication of hierarchical structures by using a surface sol–gel process. Angew Chem Int Ed 43:2746-2748

Caruso RA, Antonietti M (2001) Sol–gel nanocoating: an approach to the preparation of structured materials. Chem Mater 13:3272-3282

Caruso RA, Schattka JH, Greiner A (2001) Titanium dioxide tubes from sol–gel coating of electrospun polymer fibers. Adv Mater 13:1577-1579

Chen Q, Zhou W, Du G, Peng L (2002) Trititanate nanotubes made via a single alkali treatment. Adv Mater 14:1208-1211

Chen RJ, Bangsaruntip S, Drouvalakis KA, Kam NWS, Shim M, Li Y, Kim W, Utz PJ, Dai H (2003) Noncovalent functionalization of carbon nanotubes for highly specific electronic biosensors. Proc Natl Acad Sci USA 100:4984-4989

Chia S, Urano J, Tamanoi F, Dunn B, Zink JI (2000) Patterned hexagonal arrays of living vells in sol–gel silica films. J Am Chem Soc 122:6488-6489

Cook G, Timms PL, Göltner-Spickermann C (2003) Exact replication of biological structures by chemical vapor deposition of silica. Angew Chem Int Ed 42:557-559

Cui Y, Wei Q, Park H, Lieber CM (2001) Nanowire nanosensors for highly sensitive and selective detection of biological and chemical species. Science 293:1289-1292

Davis SA, Burkett SL, Mendelson NH, Mann S (1997) Bacterial templating of ordered macrostructures in silica and silica-surfactant mesophases. Nature 385:420-423

Dong A, Wang Y, Tang Y, Ren N, Zhang Y, Yue Y, Gao Z (2002) Zeolitic tissue through wood cell templating. Adv Mater 14:926-929

Ekanayake EMIM, Preethichandra DMG, Kaneto K (2007) Polypyrrole nanotube array sensor for enhanced adsorption of glucose oxidase in glucose biosensors. Biosens Bioelectron 23:107-113

El-Zahab B, Jia H, Wang P (2004) Enabling multienzyme biocatalysis using nanoporous materials. Biotechnol Bioeng 87:178-183

Ercolessi F, Andreoni W, Tosatti E (1991) Melting of small gold particles: mechanism and size effects. Phys Rev Lett 66:911-914

Fullam S, Cottell D, Rensmo H, Fitzmaurice D (2000) Carbon nanotube templated self-assembly and thermal processing of gold nanowires. Adv Mater 12:1430-1432

Gu Y, Huang J (2009) Fabrication of natural cellulose substance derived hierarchical polymeric materials. J Mater Chem 19:3764-3770

Hall SR, Bolger H, Mann S (2003) Morphosynthesis of complex inorganic forms using pollen grain templates. Chem Commun: 2784-2785

Han R, Xing X, Wang Y, Long Y, Sun Y, Zhao Z, Mi H (2008) Separation/enrichment of active natural low content protein using protein imprinted polymer. J Chromatogr B 873:113-118

Hassani A, Dupuis A, Skorobogatiy M (2008) Porous polymer fibers for low-loss Terahertz guiding. Opt Express 16:6340-6351

Huang J, Ichinose I, Kunitake T (2002a) Replication of dendrimer monolayer as nanopores in titania ultrathin film. Chem Commun:2070-2071

Huang J, Ichinose I, Kunitake T (2005a) Nanocoating of natural cellulose fibers with conjugated polymer: hierarchical polypyrrole composite materials. Chem Commun: 1717-1719

Huang J, Ichinose I, Kunitake T (2006a) Biomolecular modification of hierarchical cellulose fibers through titania nanocoating. Angew Chem Int Ed 45:2883-2886

Huang J, Ichinose I, Kunitake T, Nakao A (2002b) Preparation of nanoporous titania films by surface sol–gel process accompanied by low-temperature oxygen plasma treatment. Langmuir 18:9048-9053

Huang J, Ichinose I, Kunitake T, Nakao A (2002c) Zirconia-titania nanofilm with composition gradient. Nano Lett 2:669-672

Huang J, Kaner RB (2004) A general chemical route to polyaniline nanofibers. J Am Chem Soc 126:851-855

Huang J, Kunitake T (2003) Nano-precision replication of natural cellulosic substances by metal oxides. J Am Chem Soc 125:11834-11835

Huang J, Kunitake T, Onoue S (2004) A facile route to a highly stabilized hierarchical hybrid of titania nanotube and gold nanoparticle. Chem Commun, p 1008-1009

Huang J, Matsunaga N, Shimanoe K, Yamazoe N, Kunitake T (2005b) Nanotubular SnO_2 templated by cellulose fibers: synthesis and gas sensing. Chem Mater 17: 3513-3518

Huang J, Wang X, Wang Z (2006b) Controlled replication of butterfly wings for achieving tunable photonic properties. Nano Lett 6:2325-2331

Imai H, Iwaya Y, Shimizu K, Hirashima H (2000) Preparation of hollow fibers of tin oxide with and without antimony doping. Chem Lett 29:906-907

Kam NWS, Dai H (2005) Carbon nanotubes as intracellular protein transporters: generality and biological functionality. J Am Chem Soc 127:6021-6026

Kasuga T, Hiramatsu M, Hoson A, Sekino T, Niihara K (1998) Formation of titanium oxide nanotube. Langmuir 14:3160-3163

Kemell M, Pore V, Ritala M, Leskelä M, Lindén M (2005) Atomic layer deposition in nanometer-level replication of cellulosic substances and preparation of photocatalytic TiO_2/cellulose composites. J Am Chem Soc 127:14178-14179

Kim Y (2003) Small structures fabricated using ash-forming biological materials as templates. Biomacromolecules 4:908-913

Klemm D, Heublein B, Fink HP, Bohn A (2005) Cellulose: fascinating biopolymer and sustainable raw material. Angew Chem Int Ed 44:3358-3393

Kunitake T, Fujikawa S (2003) Nanocopying as a means of 3D nanofabrication: scope and prospects. Aust J Chem 56:1001-1003

Lakshmi BB, Dorhout PK, Martin CR (1997) Sol–gel template synthesis of semiconductor nanostructures. Chem Mater 9:857-862

Li Y, Cunin F, Link JR, Gao T, Betts RE, Reiver SH, Chin V, Bhatia SN, Sailor MJ (2003) Polymer replicas of photonic porous silicon for sensing and drug delivery applications. Science 299:2045-2047

Lin X, Blake AJ, Wilson C, Sun X, Champness NR, George MW, Hubberstey P, Mokaya R, Schröder M (2006) A porous framework polymer based on a Zinc(II) 4,4′-Bipyridine-2,6,2′,6′-tetracarboxylate: synthesis, structure, and “Zeolite-like” behaviors. J Am Chem Soc 128:10745-10753

Liu S, Gan L, Liu L, Zhang W, Zeng H (2002) Synthesis of single-crystalline TiO_2 nanotubes. Chem Mater 14:1391-1397

Lundqvist M, Sethson I, Jonsson BH (2004) Protein adsorption onto silica nanoparticles: conformational changes depend on the particles’ curvature and the protein stability. Langmuir 20:10639-10647

Mariano MB, Gustavo CN, Maria CM, Cesar AB (2005) Porous carbon-carbon composite replicated from a natural fibre. Chem Commun: 5896-5898

Martin CR (1994) Nanomaterials: a membrane-based synthetic approach. Science 266: 1961-1966

Meldrum FC, Seshadri R (2000) Porous gold structures through templating by echinoid skeletal plates. Chem Commun:29-30

Morris RE, Wheatley PS (2008) Gas storage in nanoporous materials. Angew Chem Int Ed 47:4966-4981

Patzke GR, Krumeich F, Nesper R (2002) Oxidic nanotubes and nanorods: anisotropic modules for a future nanotechnology. Angew Chem Int Ed 41:2446-2461

Perez GP, Crooks RM (2004) Pore-bridging poly(dimethylsiloxane) membranes as selective interfaces for vapor-phase chemical sensing. Anal Chem 76:4137-4142

Pouget E, Dujardin E, Cavalier A, Moreac A, Valéry C, March-Artzner V, Weiss T, Renault A, Paternostre M, Artzner F (2007) Hierarchical architectures by synergy between dynamical template self-assembly and biomineralization. Nature Mater 6: 434-439

Sanchezl C, Arribart H, Guille MMG (2005) Biomimetism and bioinspiration as tools for the design of innovative materials and systems. Nature Mater 4:277-288

Schattka JH, Wong EHM, Antonietti M, Caruso RA (2006) Sol–gel templating of membranes to form thick, porous titania, titania/zirconia and titania/silica films. J Mater Chem 16:1414-1420

Shan D, He Y, Wang S, Xue H, Zheng H (2006) A porous poly(acrylonitrile-co-acrylic acid) film-based glucose biosensor constructed by electrochemical entrapment. Anal Biochem 356:215-221

Shigapov AN, Graham GW, McCabe RW, Plummer HK Jr (2001) The preparation of high-surface area, thermally-stable, metal-oxide catalysts and supports by a cellulose templating approach. Appl Catal A 210:287-300

Shin Y, Li X, Wang C, Coleman JR, Exarhos GJ (2004) Synthesis of hierarchical titanium carbide from titania-coated cellulose paper. Adv Mater 16:1212-1215

Shin Y, Liu J, Chang JH, Nie Z, Exarhos GJ (2001) Hierarchically ordered ceramics through surfactant-templated sol–gel mineralization of biological cellular structures. Adv Mater 13:728-732

Tanaka D, Higuchi M, Horike S, Matsuda R, Kinoshita Y, Yanai N, Kitagawa S (2008) Storage and sorption properties of acetylene in jungle-gym-like open frameworks. Chem Asian J 3:1343-1349

Vertegel AA, Siegel RW, Dordick JS (2004) Silica nanoparticle size influences the structure and enzymatic activity of adsorbed lysozyme. Langmuir 20:6800-6807

Welbes LL, Borovik AS (2005) Confinement of metal complexes within porous hosts: development of functional materials for gas binding and catalysis. Acc Chem Res 38:765-774

Xia Y, Yang P, Sun Y, Wu Y, Mayers B, Gates B, Yin Y, Kim F, Yan H (2003) One-dimensional nanostructures: synthesis, characterization, and applications. Adv Mater 15:353-389

Xiao R, Cho SI, Liu R, Lee SB (2007) Controlled electrochemical synthesis of conductive polymer nanotube structures. J Am Chem Soc 129:4483-4489

Xu C, Tamaki J, Miura N, Yamazoe N (1991) Grain size effects on gas sensitivity of porous SnO_2-based elements. Sens Actuat B 3:147-155

Yang D, Qi L, Ma J (2002) Eggshell membrane templating of hierarchically ordered macroporous networks composed of TiO_2 tubes. Adv Mater 14:1543-1546

Zabetakis D, Dinderman M, Schoen P (2005) Metal-coated cellulose fibers for use in composites applicable to microwave technology. Adv Mater 17:734-738

Zhang X, Goux WJ, Manohar SK (2004) Synthesis of polyaniline nanofibers by "nanofiber seeding". J Am Chem Soc 126:4502-4503

Zhou J, Chang C, Zhang R, Zhang L (2007) Hydrogels prepared from unsubstituted cellulose in NaOH/urea aqueous solution. Macromol Biosci 7:804-809

5

Nanoporous Template Synthesized Nanotubes for Bio-related Applications

Yue Cui[1], **Qiang He**[1], **Junbai Li**[1, 2*]
[1] Beijing National Laboratory for Molecular Sciences (BNLMS), International Joint Lab, Institute of Chemistry, Chinese Academy of Sciences, Beijing, 100190, China.
[2*] National Center for Nanoscicence and Technology, Beijing, 100190, China.
E-mail: jbli@iccas.ac.cn

The porous template synthesis method has attracted significant interest as a versatile approach to prepare tubular nanomaterials with tailored properties. The process involves deposition or synthesis of various materials such as polymers, nanoparticles, proteins, dyes, and organic or inorganic small molecules within the porous templates, which are subsequently removed to yield free-standing nanotubes. At the same time, this approach permits the formation of composite nanotubes with the engineering features including size, shape, composition, and function. In this chapter, we summarize the synthesis and properties of various composite nanotubes based on template method combining with layer-by-layer assembly, sol–gel chemistry and polymerization. These nanotubes possess potential applications in biomedical fields such as bioseparation, biocatalysis, biosensor, and drug delivery.

5.1 Introduction

Since the first discovery of carbon nanotube in 1991 (Iijima, 1991), tubular nanomaterials have attracted great interests and have been explored as application materials due to their excellent properties and particular morphologies. Besides carbon nanotubes, the range of materials that can be formed into nanotubes has been significantly extended during the past decade, especially those with multicomponents or multifunctional properties. Among the various methods available for fabricating multi-component non-carbon nanotubes, the "porous template method" is recently known as a convenient and versatile method for producing nanotubes (Martin, 1994; Hulteen, Martin, 1997). This method involves synthesizing nanotubes with the desired materials within the uniform cylindrical pores of template membranes. Since Martin et al. first introduced the growth of one-dimensional nanomaterials in the template pores, various nanotubes or nanowires of polymers, metals, semiconductor, and other materials prepared by the template method have been widely reported (Lakshmi et al., 1997a; Steinhart et al., 2002; Lee et al., 2001; Zelenski et al., 1998; Jirage et al., 1997; Steinhart et al., 2004).

An important feature of the template method is its capability to control the dimensions and structure of the obtained nanotubes. The outside diameter of the nanotubes is determined by the diameter of the template pores, and the length is limited by the thickness of the template. With a narrow diameter distribution and nearly parallel porous structure, porous membranes such as alumina and polycarbonate (PC) are commonly used as templates in the preparation of nanotubes. These templates are both tunable with respect to length and pore diameter, allowing the dimensions of the nanotubes to be controlled precisely. The payload capacity is another issue. By comparison with the well developed nanoparticles, nanotubes have larger inner diameters, which allow nanotubes to carry a correspondingly larger payload. With differential functionalization of the inner and outer surfaces (for example, with a specific antibody on the outer surface), template synthesized nanotubes can delivery a payload (*e.g.* drug or gene) to a targeting site with high efficiency.

In this book chapter, we will review several types of processes and techniques which combine with the template method to fabricate silica and polymeric nanotubes, including layer-by-layer (LbL) assembly, sol–gel chemistry, as well as polymerization. The fundamental properties and bio-related applications of the nanotube will be discussed.

5.2 Porous Templates

Up to now, most of the work in template synthesis has employed the use of two types of nanoporous membranes: anodic alumina oxide (AAO) and "track-etch" polymeric membranes. AAO membranes are prepared via the anodization of aluminium metal in an acidic solution. These membranes contain cylindrical pores of uniform diameter arranged in a hexagonal array (Fig.5.1). Pore size, shape, and density can be varied in a controlled manner by the proper selection of the anodization conditions. The diameter value of the pore can range from several to hundreds of nanometers; whereas the pore densities as high as 10^{11} pores per centimeter can be achieved. The AAO membrane is stable at temperatures at which soft matteris commonly processed, and resistant against organic solvents; but can be selectively etched with aqueous acids and bases to release nanotubes fabricated inside its pores.

Fig.5.1. Scanning electron microscopy (SEM) images of a typical AAO membrane. a) Surface and b) cross-section, with pore diameter of approximately 70 nm. Reprinted from Hillebrenner et al., 2006. Copyright (2006), with permission from the Future Medicine Ltd

The second type of common hard templates are track-etch membranes (Fig.5.2) (Grawford et al., 1992), which are produced by irradiating polymeric films, with a thickness ranging from a few microns to a few tens of microns, with ion beams; thus producing latent tracks penetrating through the bombarded films. Next, pores are generated at the positions of the latent tracks by wet-chemical etching. Pore size, shape, and density can be varied in a controllable manner by the proper selection of the conditions under which irradiation and post-treatment procedures are carried out. Pores with diameter values ranging from 10 nm to the micron range are obtained; whereas, the pore density can be adjusted to any value between 1 to 10^{10} pores per centimeter. Moreover, diameter value and pore density can be adjusted independently. The most common polymers track-etch membranes are composed of polycarbonate (PC) and polyethylene terephthalate. The pore walls are commonly hydrophilized by plasma treatment, or by adsorbing, or grafting hydrophilic polymers, such as polyvinyl pyrrolidone, onto the pore walls. The two main limitations associated with track-etch membranes are their limited stability at elevated temperatures and their poor resistance against organic solvents, which poses problems for many of the self-assembly processes. The arrangement of the pores is random, that is, track-etch membranes do not exhibit long-range order. Moreover, because of their poor rigidity and their lack of chemical resistance to organic solvents, it is difficult to remove residual material from the surface of track-etch membranes after their infiltration; a process step that is crucial to the template-based fabrication of nanotubes and nanorods. Nevertheless, because of their commercial availability and versatility, track-etch membranes are being routinely used for the production of one-dimensional nanostructures. However, it was found that the pronounced roughness of the pore walls in track-etch membranes, revealed by scanning electron microscopy (SEM) and adsorption experiments, prevents uniform orientation of anisotropic species infiltrated into the pores (Steinhart, 2008).

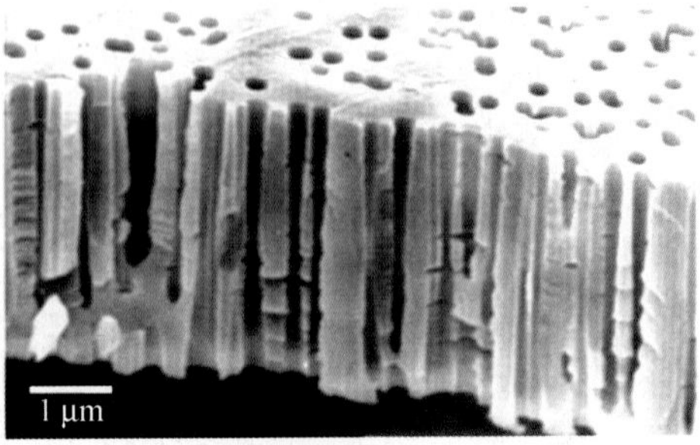

Fig.5.2. Example of a polymeric track-etch membrane. Reprinted from Steinhart, 2008. Copyright (2008), with permission from Springer

5.3 Preparation of Composite Nanotubes in Porous Template

Many types of strategies have been used to prepare template-synthesized nanotubes, such as LbL assembly, sol–gel chemistry, and polymerization. These processes involve synthesizing desired materials within the pores of a porous membrane. Depending on the properties of the materials and the chemistry of the pore wall, nanotubes with different properties and applications can be obtained.

5.3.1 LbL-assembled Polymeric Nanotubes

The LbL assembly technique, which was introduced by Decher (1997) and colleagues, was initially based on alternating deposition of polyelectrolytes, with opposite charges, on a planar substrate. Up to now, the assembly driving forces have been extended to covalent bond, hydrogen bond, base pair interaction, and host-guest interaction. The LbL technique allows the coating of diverse species in various shapes and sizes, with uniform layers and controllable thickness (Donath et al., 1998; Caruso, 2000; Peyratout, Daehne, 2004). It has led to a number of advances in material science, ranging from the development of novel optical and electronic properties, and the formation of high strength materials, which mimic nature, to stimuli-responsive materials (Hammond, 2004; Duan et al., 2007a; Wang et al., 2007). The process of the preparation of composite nanotubes via LbL technique is simple and versatile. Briefly, a piece of porous membrane template, such as AAO or PC, is first immersed into a solution containing component A for a certain time. The templates were then rinsed with a suitable solvent three times in different beakers. Next, the component B was alternately adsorbed in the pores of the membrane and then washed three times. This cycle can be repeated until the desired number of layers are obtained. After the multiple layers that had been deposited on the top and bottom surface of the membrane were removed, the nanotubes are released by dissolving the template under certain conditions. Herein, we describe the preparation and properties according to their interaction between different components.

5.3.1.1 Electrostatic LbL-assembled Nanotubes

Electrostatic interaction is the most common driving force to fabricate polyelectrolyte multilayers. Poly(allylamine hydrochloride) (PAH)/poly(sodium styrenesulfonic) (PSS) are a classic assembly pair which has been widely applied to prepare polyelectrolyte multilayer films on the planar and colloid templates. After the removal of the substrates or templates, the free-standing PAH/PSS multilayer or hollow PAH/PSS multilayer capsules can be released. The properties of PAH/PSS multilayer film have been investigated in detail with regard to temperature, salt concentration, pH, and mechanical strength. Readers can reference several good reviews (Caruso, 2000; Peyratout et al., 2004; Hammond et al., 2004; Wang et al., 2008; He et al., 2009).

The combination of the electrostatic LbL technique with the porous template method was initially developed so that polymeric tubular structures with complex, but well-controlled, wall morphologies and adjustable wall thickness can be prepared. In order to obtain continuous tubes and avoid clogging the template, a so-called pressure-filter-template method was proposed by our group in 2003 (Scheme 5.1) (Ai et al., 2003). In this way, one can overcome the difficulty by allowing the components to smoothly filter through and be completely deposited along the pore wall of the template. With this method, we prepared polyelectrolyte nanotubes through the LbL adsorption of PSS and PAH in the inner wall surface of the AAO membrane. Fig.5.3 shows the SEM images of the regular polyelectrolyte nanotube arrays after the complete removal of the template. The tube wall consists of three PAH/PSS layers, and the wall thickness is 50~80 nm and the length is up to 60 μm, the order of the AAO template thickness. However, the thickness of the nanotube walls was one order of magnitude larger than that of corresponding multilayer structures prepared on smooth substrates, in which a bilayer has a thickness of several nanometers. Caruso and coworkers deposited poly(ethylenimine) (PEI), poly(acrylic acid) (PAA)/PAH multilayers onto the pores of PC membranes, with a diameter of 400 nm, in the presence of Cu^{2+} and then thermally cross-linked them (Liang et al., 2003). Whereas, the wall thickness of the nanotubes obtained could be adjusted by the number of successive deposition cycles, the functionality of the embedded inorganic nanoparticles was preserved. The wall thicknesses of the nanotubes reported in this study were only slightly larger than those in smooth configurations.

Lee and coworkers (2006) also prepared PAH/PSS nanotube membranes at neutral conditions by using the same method. At a high pH condition (pH > 9.0), the PAH/PSS multilayers in the PC membrane behave differently causing discontinuous swelling/deswelling transitions, which leads to a pH deduced hysteretic gating property of the membrane. The flux of pH-adjusted water was used to detect the transitions. It showed that the PAH/PSS multilayers in the confined geometry swelled to smaller extents compared to the same multilayers on planar substrates under the same conditions. And the average thickness of a bilayer in the cylindrical pores of PC membrane is

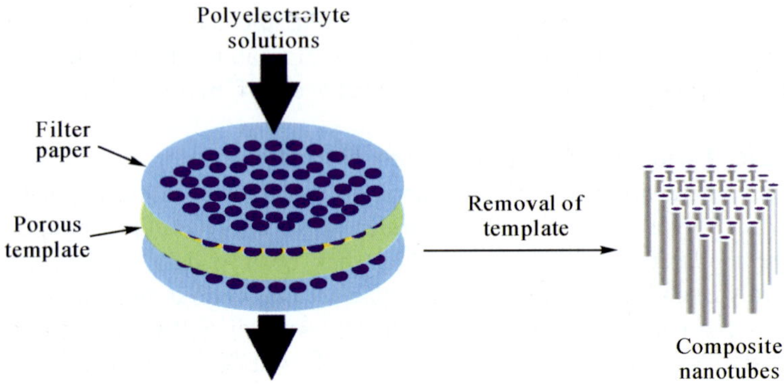

Scheme 5.1. Scheme of the pressure-filter-template method

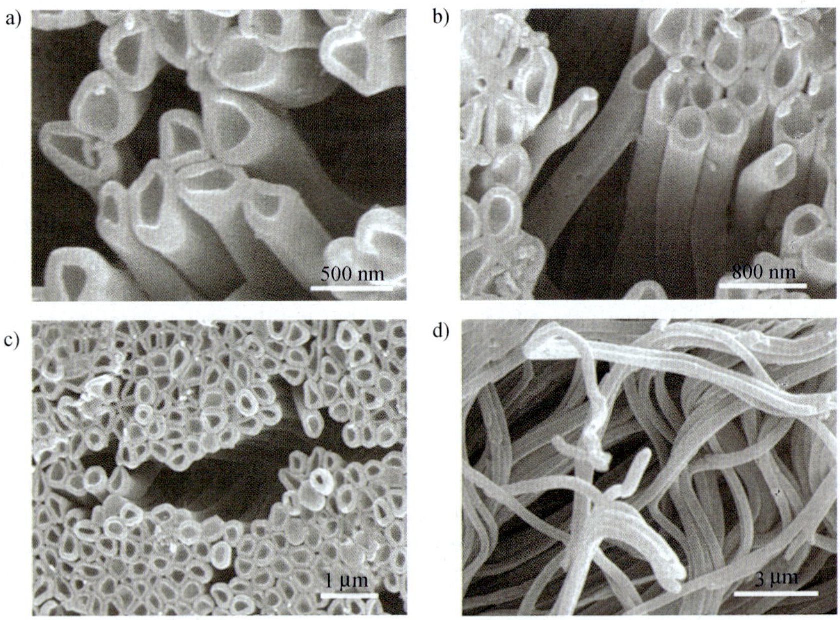

Fig.5.3. Scanning electron microscopy (SEM) micrographs of polyelectrolyte nanotubes via electrostatic LbL assembly. a) High magnitude SEM image of the polyelectrolyte tubes through the PAH/PSS assembly. b) PAH/PSS nanotubes with very thick wall structure and smooth surface. c) Ordered array of polymer tubes after the complete removal of the alumina template. d) Highly flexible tubes. Reprinted from Ai et al., 2003. Copyright (2003), with the permission from ACS

greater than that of planar substrates. The hysteretic gating property of the multilayer-modified PC membrane was utilized to achieve either a "closed" or "open" state at one pH condition, depending on the pretreatment history. This gating property enables either the retention or passage of high-molecular weight polymers by varying the membrane pretreatment condition as a stimuli-responsive chemical valve.

Polyelectrolyte multilayers have been proven to be excellent hosts for electrical, optical, and electrochemically active materials systems; therefore, various functional nanotubes with multilayer films can be obtained after removal of the template. In a report of our group, the conductive polymer, a negatively charged polypyrrole (PPy), was used to fabricate conductive nanotubes through the alternating adsorption with positively charged PAH on PC pore's surfaces (Ai et al., 2005). The assembled PPy/PAH nanotubes are rather stable without the requirement of adjusting pH values to provide charge density. The obtained nanotube replicated the shape of the template with a diameter of 400 nm and a length of 10 μm. The electro-property of the assembled nanotubes has been characterized by the cyclic voltammetry (CV) measurement. It shows that PPy has a stable oxidation state in organic acid. From the alternating current impendence measurement, the PPy/PAH nanotubes conductivity is 0.008 $S \cdot cm^{-1}$, which agrees well with the results for the PPy/PAH microcapsules or films on a flat substrate, indicating the conductive property of the assembled nanotubes (Zheng et al., 2004). The mechanical stability of as-prepared nanotubes depended on the number of the assembled bilayers. Six PPy/PAH bilayers were observed as a critical condition for constructing stable nanotubes, while below six the nanotubes will collapse.

There is increasing interest in the fabrication of composite nanomaterials such as zero-dimensional or one dimensional nanostructures consisting of semiconductor, metal, or organic nanoparticles. If the composites possess specific structures and remarkable optical, electrical, magnetic, or chemical properties, the assembled composite nanomaterials will exhibit the similar features (Gao et al., 1999; Willner et al., 2001; Wang, Chumanov, 2003). Liang and co-workers (2003) reported the assembly of polyelectrolyte/nanoparticle hybrid nanotubes. To prepare this kind of nanotubes, a multilayer polyelectrolyte film (for example, PEI/PSS/PAH/PSS) should be first deposited onto the surface of the template. This primer film can reduce the influence of the membrane on nanoparticle adsorption and provide a uniformly charged surface to facilitate the adsorption of the reverse-charged nanoparticles. The nanoparticles are then adsorbed with alternating polyelectrolyte multilayers to yield membrane-supported coatings of $(\text{polyelectrolyte/nanoparticle})_n$. In their experiments, Au and CdTe nanoparticles are successfully introduced into nanotubes. Fig.5.4a shows a transmission electron microscopy (TEM) image of a $[(\text{PEI/PSS/PAH})/\text{CdTe}]_6$ nanotube, and a high-resolution TEM image shows that CdTe nanoparticles are present in the nanotube (Fig.5.4a inset). A confocal laser scanning microscopy (CLSM) image of the $[(\text{PEI/PSS/PAH})/\text{CdTe}]_6$ nanotubes further confirms the tubular structure and shows luminescence of the CdTe nanoparticles present in the nanotubes (Fig.5.4b). The spectrums results also prove the existence of the nanoparticles.

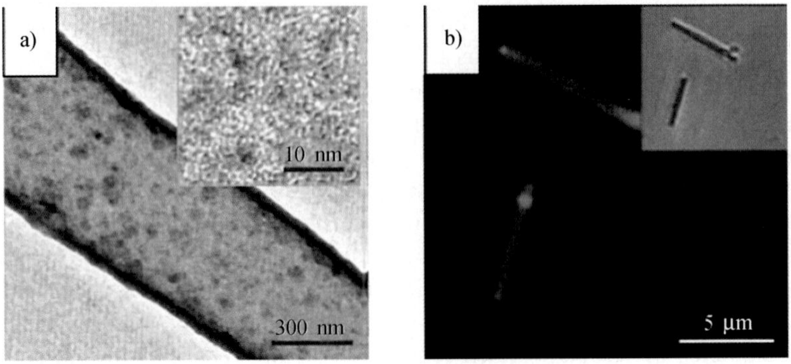

Fig.5.4. a) Transmission electron microscopy (TEM) image of a [(PEI/PSS/PAH)/CdTe]$_6$ nanotube, the inset is a high-resolution TEM image of tube wall. b) Confocal microscopy image of [(PEI/PSS/PAH)/CdTeNP]$_6$ nanotubes, the inset shows the corresponding transmission mode. Reprinted from Liang et al., 2003. Copyright (2003), with permission from Wiley

Interestingly, we found that the phenomenon of Rayleigh instability occurs when (PSS/PAH) multilayer nanotubes are hydrothermally treated above 121 °C (He et al., 2008a). It is well-known that a fluid cylinder with a circular cross-section breaks up into small spherical droplets with the same volume, but less surface area, if its length exceeds its circumference. The Rayleigh instability is driven by the surface tension of liquids to decrease surface tension or to minimize their surface areas. Hydrothermal annealing of the (PAH/PSS)$_8$/PAH nanotubes, at temperatures above the glass transition temperature, caused the growth of polyelectrolyte multilayer thickness. Subsequently the formation of a pearl-necklace-like structure, and finally the structural transformation of (PAH/PSS)$_8$/PAH nanotubes from nanotubes to capsules occurs. The diameter of the obtained capsules is in a range of 500 nm to 1.00 μm, which is obviously larger than that of the initial tubes (~400 nm). However, their sizes are less than the theoretical value (~1.5 μm) according to Rayleigh instability. This was ascribed to the shrinkage of polyelectrolyte multilayers during the transformation process. In addition, the amount of the assembled polyelectrolyte layers has a significant influence on the transformation process. The structural transformation of polyelectrolyte multilayers from tubes to capsules after annealing was explained by the input of thermal energy, which leads to a breakage of ion pairs between oppositely charged polyelectrolyte groups. The shrinkage of the obtained capsules is accompanied by a strong increase of wall thickness and a smoothing of the surface. The driving force of this rearrangement process is caused by the entropy increase through the more coiled state of the polyelectrolyte molecules and the decrease of interface. This structural transformation offers an elegant and perspective approach for manufacturing polymer nanomaterials in a controlled manner, which can have potential applications as delivery devices.

5.3.1.2 LbL-assembled Nanotubes via Hydrogen Bonding

Initially, LbL assembly was focused primarily on the use of commercially available polyelectrolytes for constructing multilayer films based on electrostatic interaction. However, subsequent work has demonstrated that, in addition to electrostatic attraction, a number of different driving forces for multilayer buildup can be exploited (Quinn et al., 2007). Such versatility of the driving force means that the LbL method is not restricted to charged materials and water solutions. This breakthrough is of critical importance as many functional polymers are uncharged, and therefore, electrostatic-driving LbL assembly does not permit the formation of thin films from these polymers. One of the most commonly studied, non-electrostatic interactions used in LbL assembly to date is hydrogen bonding. By exploiting this interaction, a host of different materials have been successfully incorporated into multilayer films in a water solution or organic phase (Stockto et al., 1997; Wang et al., 1997; Sukhishvili et al., 2000; Sukhishvili et al., 2002; Yang et al., 2002; Quinn et al., 2004; Zelikin et al., 2006). This possibility arises because many polymers incorporate moieties that can act as hydrogen bonding donors and acceptors. For instance, the oxygen atoms in the carboxylic acid groups of poly(acrylic acid) (PAA) molecules can be hydrogen bonding acceptors; and donors (nitrogen) in the pyridine rings of poly(4-vinyl-pyridine) (PVP) molecules are also present (Wang et al., 1997). Based on these studies, we fabricated PAA/PVP nanotubes in the PC template through a hydrogen bonding LbL technique in a methanol solution (Fig.5.5a). The regular $(PAA/PVP)_5$ array exhibits tubes with smooth and clean surfaces, a wall thickness of around (50±5) nm, and length in the order of the thickness of PC membrane (*ca.* 13 μm). The fabricated $(PAA/PVP)_5$ nanotubes exhibit good stability and flexibility. As shown in Figs.5.5b and 5.5c, UV spectra results proved that the wall thickness of the nanotubes is strongly dependent on the number of PAA/PVP pairs assembled. This report of LbL-assembled composite nanotube via hydrogen bonding extends the research on the fabrication and application of composite nanotubes. Thus, a number of natural and synthetic polymers with biocompatibility and biodegradability can be readily incorporated into the walls of nanotubes, even if they cannot be assembled via traditional electrostatic LbL techniques. Importantly, hydrogen bonding provides opportunity to render films responsive to different chemical and physical stimuli, allowing the preparation of so-called “intelligent” nanotubular materials. We displayed an example for the possibility of pH-tunable film disassembly of as-prepared PAA/PVP nanotubes. After releasing PAA in a basic aqueous solution, porous nanotubes can be obtained. The pore size can be tuned by the immersion time or the pH values of the basic aqueous solution. The assembled PVP nanotubes with porous walls were stable at room temperature; and they may be applied as carriers of catalysts and drugs to achieve a better dispersion and diffusion of species, especially in an aqueous system.

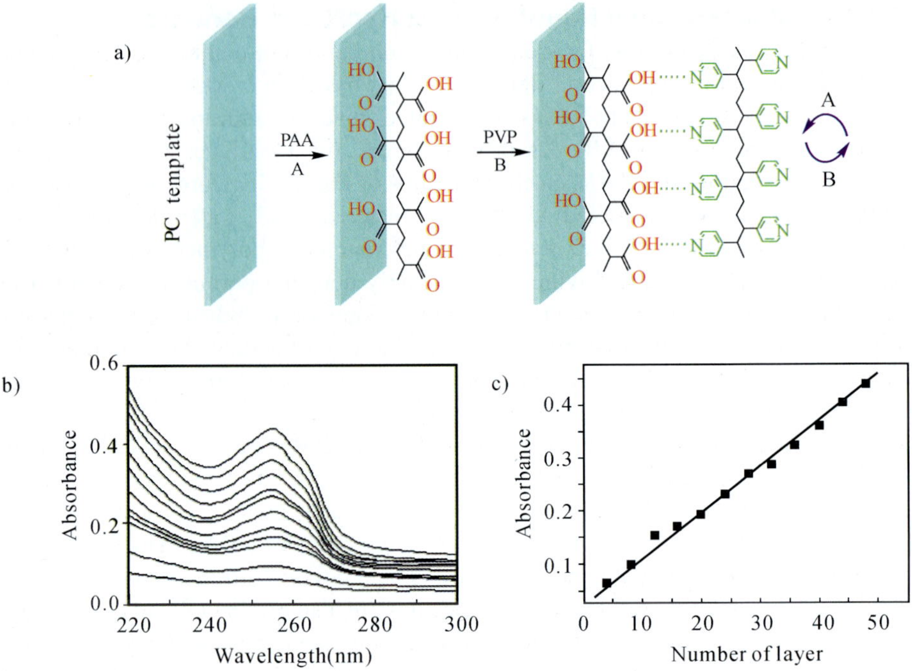

Fig.5.5. a) Assembly of PAA/PVP nanotubes within the walls of the PC template by means of the LbL technique based on hydrogen bonding. b) UV spectra of $(PAA/PVP)_n$ multilayer films, in which n=4~48, assembled on a quartz substrate. The spectra were obtained after every four cycles of assembly. c) Features absorbance of pyridine from PVP at 256 nm versus the number of layers deposited. Reprinted from Tian et al., 2006b. Copyright (2006), with permission from Wiley

An extension of hydrogen-bonded LbL multilayer films is utilizing the base pairing of DNA nucleotides to assist film assembly. Hybridization of DNA to form a double helix occurs naturally, where the base adenosine (A) pairs with thymidine (T) and cytosine (C) pairs with guanine (G). The driving force for forming double stranded (ds) DNA is a combination of the hydrogen bonds between the bases and π-π stacking of the aromatic rings contained in the bases. As the formation of dsDNA is dependent on the correct recognition of base pairs, the structure of the DNA multilayer film can be manipulated by altering the sequence of bases. Actually, DNA has been used as a polyelectrolyte to be incorporated into the multilayer film via electrostatic interactions in numerous systems (Peyratout, Daehne, 2004; Quinn et al., 2007). However, this approach does not utilize the special interactions between base pairs which can be used to finely engineer the structure of the multilayer film (Johnston et al., 2005; 2006). Additionally, DNA is biocompatible and biodegradable, which makes it an attractive building block for forming multilayer films. Hou et al. (2005a) have reported the first example of DNA nanotubes by the sequential deposition of complementary oligonucleotides with specific sequences in the AAO membrane. Both faces of the membrane were first sputtered with an about 5 nm gold film which can effectively resist in the adsorption of the assembled materials. Subsequently, the nanopore alumina template is immersed into a solution of 1,10-decanediylbis(phosphonic acid) (α,ω-DOP), and then into a solution of $ZrOCl_2$, resulting in attachment of a layer of α,ω-DOP/Zr (IV) along the pore walls via the Mallouk's phosphonate chemistry. The outer α,ω-DOP/Zr (IV) layer serves as a nanotube skin which can provide structural integrity, surrounding an inner core of multiple double-stranded DNA layers held together by hybridization (*i.e.* complementary hydrogen bonding formation) between the DNA layers (Fig.5.6). The DNA molecules comprising the tubes can be varied at will, and the DNA can be released from the tube by changing the environmental conditions such as temperature.

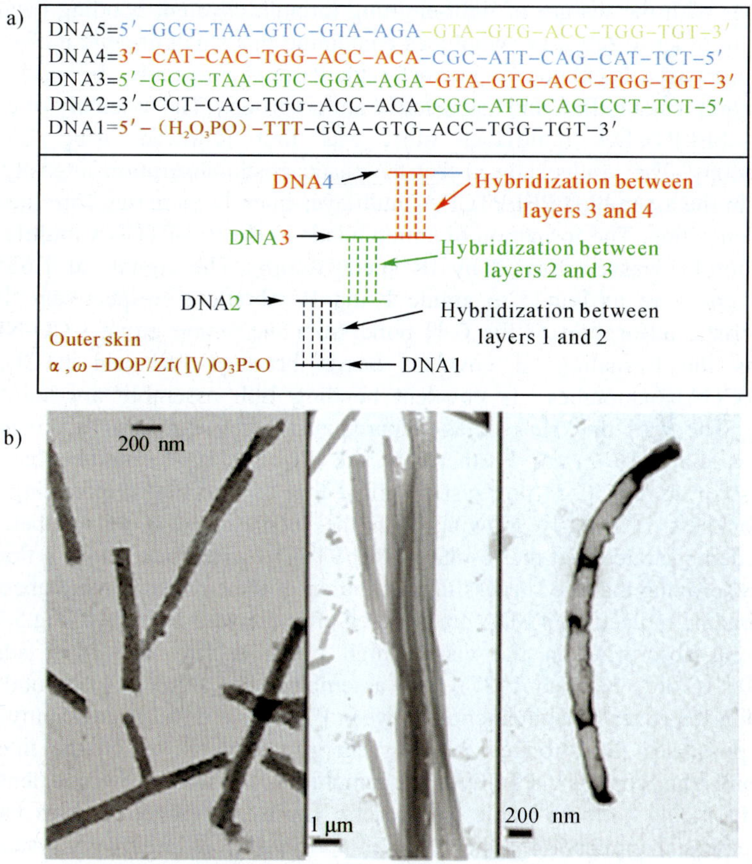

Fig.5.6. a) DNAs used to prepare nanotubes where hybridization between segments 15 bases long hold the layers together, and schematic of the hybridization-based layer-by-layer DNA nanotube concept. b) TEM images of five-layer DNA. Reprinted from Hou et al., 2005a. Copyright (2005), with permission from ACS

5.3.1.3 LbL-assembled Nanotubes Based on Covalent Bonding

Other non-charged driving forces, such as covalent bonding, are also applied to fabricate LbL multilayer films (Kohli et al., 2000; Serizawa et al., 2002; Zhi et al., 2005; Such et al., 2006; Duan et al., 2007b). Although the use of covalent bonds to assemble multilayer films via the LbL technique has been a more recent area of investigation, such films can provide significant advantages to the usual fabrication methods for some applications. In particular, they have high stability due to the covalent bonds formed. They can also be fabricated in organic solvents which is eventually impossible with electrostatic interactions, because a polycation and polyanion will form a salt and precipitate in solution. We have reported that dye molecule, 3, 4, 9, 10-perylenetetracarboxylicdianhydride (PTCDA) (Zhi et al., 2005; Djuristic et al., 2000), is one of the perylene derivatives selected to make composite nanotubes with the feature of light-emitting through covalent bonding (Tian et al., 2006c). In order to fabricate nanotubes containing this small organic molecule, PEI was chosen as a backbone to interact with PTCDA molecules because PEI can readily form a stable film and has reactive amino groups. The assembly of covalent-bonded PEI/PTCDA multilayer films was first followed using UV-visible spectroscopy. The results showed that the characteristic absorption intensity of UV spectra in the assembled PEI/PTCDA multilayer linearly increases with the number of the deposition. The formation of amide bonds in the PEI/PTCDA multilayer film and nanotube was confirmed by IR spectroscopy. The signals at 1,635.5 and 1,566.1 cm^{-1} are attributed to amide I and II vibrations, respectively. It is the characteristic adsorption of the C-N bond from the amide group (-CO-NH-) and confirms the formation of covalent bonds between PEI and PTCDA. The $(PEI/PTCDA)_6$ nanotubes via covalent bonding LbL assembly are uniform and flexible. The wall thickness of as-prepared nanotubes is (100 ± 10) nm and each bilayer is about (16 ± 2) nm. Furthermore, the $(PEI/PTCDA)_3$ nanotubes have a wall thickness of about (50 ± 5) nm, just a half of that of six circles, indicating that the wall thickness is linearly growing with the increasingly cycle number. CLSM images demonstrated that the resulting $(PEI/PTCDA)_6$ nanotubes have a fluorescent property, proving that the successful preparation of the composite nanotubes and the small organic molecule PTCDA are retained after the wall assembly (Fig.5.7). With the strong absorption in the visible high photo-stability and ideal adsorption properties (Ford, Kamat, 1987), the assembled PTCDA/PEI nanotubes are a promising sensitizer. The interaction between PTCDA and PEI is covalently bonded, which possesses the strongest binding energy compared to charge interaction. Therefore, the complex PTCDA/PEI nanotubes fabricated by covalently LbL assembly should be more stable than others. These fluorescent complex nanotubes have a potential application in the design of optical devices or delivery systems.

The advantage of covalent bonding LbL deposition lies in the fact that solutions in organic solvents can be used to deposit the monolayers. To this end, nanotubes consisting of PEI/poly(styrene-alt-maleic anhydride) (PSMA) multilayers were obtained by connecting the alternating monolayers by covalent bonds (Tian et al., 2006d). The formation of amide bonds between PEI and PSMA enables the nanotubes with a good mechanical stability. The composite nanotubes were prepared in organic solvents, which results in the as-prepared nanotubes to be used in a wide range of systems. The results demonstrate that the tubular structure of $(PEI/PSMA)_5$ is stable and has a uniform surface; however, the pure PSMA nanotubes were collapsed due to bad tube structure. Thus, the addition of PEI is necessary in order to get a complex uniform tube. Such assembled nanotubes can be considered as carriers to delivery materials on a larger scale.

When the LbL assembly is applied to synthesize nanotubes in templates, it will extend the preparation methods of nanotubes, providing a viable and facile approach, with controllable diameter and length, and containing diverse compositions. The obtained functional polymer nanotubes, through LbL assembly based on no-charged interaction, are stable at room temperature; and proved further that their ordinary properties in the multilayer can be preserved in the structure of nanotubes as that in a single polymer. This brings out new functions for the composite nanotubes and broadens the species that can be incorporated into the nanotube wall materials. These endeavors will be a good foundation for the future application of nanomaterials.

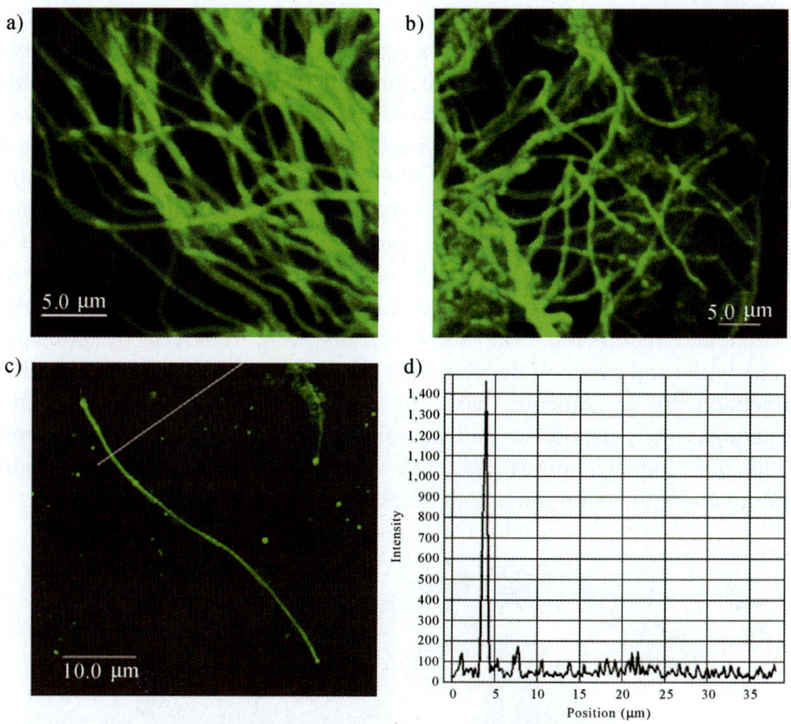

Fig.5.7. Slim and bright lines representing the $(PEI/PTCDA)_6$ nanotubes. a), b) Exhibiting the fluorescent nanotubes wrapped around different directions. c) Selected single nanotube was used for the intensity measurement. The tube length is similar to the thickness of the template membrane, 60 μm. d) Intensity profile of a fluorescent nanotube. Reprinted from Tian et al., 2006c. Copyright (2006), with permission from ACS

5.3.2 Nanotubes Bases on Sol–gel Chemistry

Sol–gel chemistry has been proven to be a powerful approach for preparing inorganic materials such as glasses and ceramics (Hench, West, 1990). In sol–gel synthesis, a soluble precursor molecule is hydrolyzed to form a dispersion of colloidal particles (the sol). Further reaction causes bonds to form between the sol particles resulting in an infinite network of particles (the gel). The gel is then typically heated to yield the desired material (Livage et al., 1988). Using this method to synthesize inorganic materials has a number of advantages over more conventional synthetic procedures. For example, high-purity materials can be synthesized at a lower temperature. In addition, homogeneous multicomponent systems can be obtained by mixing precursor solutions; this allows for easy chemical doping of the materials prepared. Finally, the rheological properties of the sol and the gel can be utilized in processing the material, for example, by dip coating of thin films, spinning of fibers, *etc.* (Aegerter et al., 1996).

Martin's group have combined the concepts of sol–gel synthesis and template preparation of nanomaterials to yield a new general route for preparing nanostructures of semiconductors and other inorganic materials. This was accomplished by conducting sol–gel synthesis within the pores of various micro- and nanoporous membranes. Monodisperse tubules and fibrils of the desired material are obtained. Micro- and nanostructures of inorganic oxides, such as TiO_2, MnO_2, V_2O_5, Co_3O_4, ZnO, WO_3, and SiO_2, have been prepared (Lakshmi et al., 1997b). For example, titanium isopropoxide was used as the precursor molecule for the sol–gel preparation of the TiO_2 nanotubes. Briefly, the AAO template is dipped into the TiO_2 sol for the desired amount of time, removed and allowed to dry in air. The sol-containing membrane is then heated in air at 400 °C for a certain time. This procedure yields TiO_2 tubes and fibrils within the pores and TiO_2 films on both faces of the template membrane. The surface films were removed by polishing the membrane with sand paper. If desired, the template membrane can then be dissolved by immersion in an aqueous base, to expose the template-synthesized TiO_2 nanostructures. TiO_2 nanotubes and nanofibrils are prepared in AAO membranes with 200 nm diameter pores (Fig.5.8). It is proven that through control of temperature and immersion time, both tubes and fibrils can be prepared.

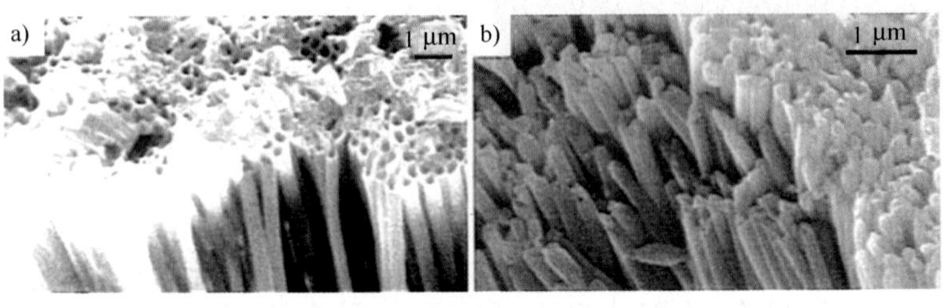

Fig.5.8. SEM images of TiO_2 nanotubes obtained by immersing the template membrane in the sol for a) 5 s and b) 25 s. Reprinted from Lakshmi et al., 1997b. Copyright (1997), with permission from ACS

The mechanism of formation of TiO_2 from acidified titanium alkoxide solutions is well discussed (Hench, West, 1990; Livage et al., 1988). In the early stages, sol particles held together by a network of -Ti-O- bonds are obtained. These particles ultimately coalesce to form a three-dimensional gel network. The tubule structures are initially obtained when this process is done in the AAO membrane, which indicates that the sol particles adsorb to the pore walls. It is well-known that, at the pH values used here, the sol particles are weakly positively charged (Bischoff, Anderson, 1995). The tubules are formed because these positively charged particles interact with anionic sites on the alumina pore wall. It is notable that the interaction of the sol particles with the AAO pore wall accelerates the rate of the gelation process. When a lower concentration of titanium was used to make the sol, gelation in the bulk solution was extremely slow, even at room temperature. However, when the alumina membrane was dipped into the TiO_2 sol, solid fibrils of TiO_2 are obtained in the pores, even at short (5 s) immersion times. These results show that gelation occurs within the pores under conditions where gelation in bulk solution is negligibly slow.

To have control over the quality of sol–gel synthesis nanotubes with more precision, a so-called "surface sol–gel" (SSG) method was used to prepare SiO_2 nanotubes (Kovtyukhova et al., 2003). This method involves repeats of two-step deposition cycles, in which the adsorption of a molecular precursor and the hydrolysis (in the case of oxide film growth) steps are separated by a post-adsorption wash. The washing step desorbs weakly bound molecules that form additional layers (Ichinose et al., 1997). In their experiment, the thickness (2~30 nm) and porosity of the SiO_2 nanotubes can be precisely controlled by varying the composition of the precursor solution and the number of adsorption/hydrolysis cycles (Scheme 5.2). Thus, both

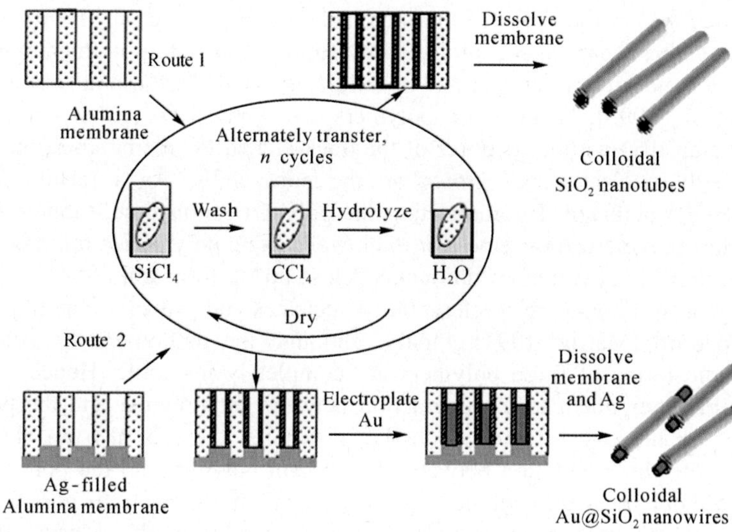

Scheme 5.2. Scheme showing the SSG synthesis of SiO_2 nanotubes (route 1) and SiO_2-coated nanowires (route 2). Reprinted from Kovtyukhova et al., 2003. Copyright (2003), with permission from Wiley

robust and flexible silica nanotubes can be obtained. The thickness of the SiO_2 layer deposited in each cycle, which always exceeds that of a monomolecular layer, can be explained by assuming occlusion of the water present as a surface layer. This phenomenon was also found for planar oxide SSG films (Ichinose et al., 1997; Kovtyukhova et al., 2000). While still in the membrane, the SiO_2-coated pores can be electrochemically filled with metal and then released by etching in acid to give free-standing insulated wires.

5.3.3 Nanotubes Synthesized by Polymerization

Polymerization provides a special way to synthesize polymer nanotubes, especially for functional and responsive polymers that are sensitive to the environment, such as electrical or thermal characterization, pH, or solvent. Chemical template synthesis of a polymer can be accomplished by simply immersing the membrane into a solution containing the desired monomer and a polymerization reagent. Martin and coworkers pioneered a process to synthesize a variety of conductive polymers within the pores of various template membranes (Cai et al., 1991; Lei et al., 1992; Parthasarathy et al., 1994a). Their work mainly focused on the electronically conductive polymers PPy, poly(3-methylthiophene), and polyaniline. These polymers can be synthesized by oxidative polymerization of the corresponding monomer, which may be accomplished either electrochemically or with a chemical oxidizing agent. The most common way to do electrochemical template synthesis is to coat a metal (*e.g.* Au) film onto one surface of the template membrane (*e.g.* PC membrane) and then use the metal film as an anode to electrochemically synthesize the polymer within the pores of the membrane (Van Dyke et al., 1990). Chemical template synthesis can be accomplished by simply immersing the membrane into a solution of the desired monomer and its oxidizing agent (Parthasarathy et al., 1994b; Martin et al., 1993). When these polymers are synthesized (either chemically or electrochemically) within the pores of the track-etched PC membranes, the polymer preferentially nucleates and grows on the pore walls. As a result, polymeric nanotubes are obtained. By controlling the polymerization time, tubules with thin walls (short polymerization times) or thick walls (long polymerization times) can be produced. For PPy, the tubules ultimately "close up" to form solid fibrils.

The reason the polymer preferentially nucleates and grows on the pore walls is straightforward (Martin, 1991). Firstly, although the monomers are soluble, the polycationic forms of these polymers are completely insoluble. Hence, there is a solvophobic component to the interaction between the polymer and the pore wall. Secondly, in the case of the conductive polymers, there is also an electrostatic component to the interaction between the nascent polymer and the pore wall. It is because the polymers are cationic, and there are anionic sites on the pore walls. This illustrates an important point: if a "molecular anchor" (Brumlik, Martin, 1991) that interacts with the material being deposited is present on the pore wall, a hollow tubule (as opposed to a solid fibril) will be obtained (Fig.5.9a). For example, Martin's group reported that Au nanotubes can be electrochemically deposited into the pores of AAO membranes while a silane that contains a -CN functionality is first bonded to the alumina pore wall (Brumlik, Martin, 1991). If the -CN containing silane is not attached to the pore wall before the Au deposition, solid Au fibers are obtained in the pores. The molecular anchor concept provides a general route for template synthesis of tubular structures (Brumlik et al., 1994; Nishizawa et al., 1995).

It is notable that the template synthesized polymer nanotubes or nanofibrils have enhanced conductivity. A plot of conductivity versus diameter for template-synthesized PPy fibrils is shown in Fig.5.9b. Whereas the fibrils with large diameters have conductivities comparable to those of bulk samples of PPy, the conductivities of nanofibrils, which have the smallest diameters, are more than an order of magnitude greater. Analogous enhancements in conductivity have been observed for template-synthesized polyaniline (Parthasarathy et al., 1994b; Martin et al., 1993) and poly(3-methylthiophene) (Cai, Martin, 1989). The template-synthesized materials have higher conductivities because the polymer chains on the outer surface of the tubules or fibrils are aligned. This can be proven with a technique called polarized infrared absorption spectroscopy (Cai et al., 1991; Liang, Martin, 1990; Lei et al., 1992).

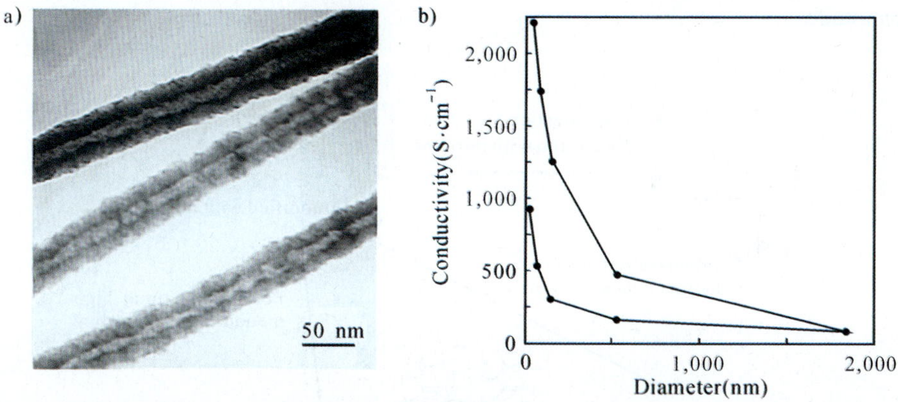

Fig.5.9. a) TEM image of polypyrrole nanotubes. b) Conductivity versus diameter for polypyrrole fibrils. Data for two different synthesis temperatures are shown: lower curve, 0 °C; upper curve, −20 °C. Reprinted from Brumlik, Martin, 1991. Copyright (1991), with permission from ACS

Among the controlled polymerization processes, atom transfer radical polymerization (ATRP) allows the synthesis of various polymers or copolymers with well-controlled and narrow molecular weight distribution, as well as defined topology (Patten, Maytjaszewski, 1998). Surface-initiated ATRP, wherein the initiator is grafted onto the surface of a substrate, provides the possibility to synthesize the desired polymer on the substrate surface through covalent bonding (Sun et al., 2004; Lee et al., 2004; Kong et al., 2004; Fu et al., 2004). Recently, it has been reported that PNIPAM-coated carbon nanotubes, which possess a thermosensitive characteristics, can be fabricated by surface-initiated ATRP (Kong et al., 2004).However, pure thermosensitive polymer nanotubes with a higher mechanical stability are of specific interest as catalyst carriers or for drug delivery. We reported the preparation of thermosensitive pure polymer nanotubes using an AAO membrane by surface-initiated ATRP via the "grafting from" strategy (Cui et al., 2005; Cui et al., 2006). The membrane pores were modified with aminopropylsilane before immobilization of the ATRP initiator, 2-bromoisobutyryl bromide, on the silanizated AAO template. A copolymer of *N*-isopropylacrylamide and *N,N'*-methylenebisacrylamide (PNIPAM-co-MBAA) was then synthesized within the membrane, the scheme is shown in Scheme 5.3. The highly flexible nanotubes could then be obtained after removal of the template. Fig.5.10 shows typical SEM images of three nanotube samples with different monomer concentrations. The maximum lengths of the nanotubes are up to several tens of micrometers, corresponding to the thickness of the AAO membrane (60 μm) used. This proved that the polymer synthesized by ATRP is able to coat the entire membrane pore wall. All of the samples show a smooth and winding shape, which demonstrates the high flexibility of the PNIPAM-co-MBAA nanotubes. TEM, atomic force microscopy (AFM), and gel permeation chromatography (GPC) proved that increasing the monomer concentration in polymerization led to a proportional increment of thickness of nanotubes.

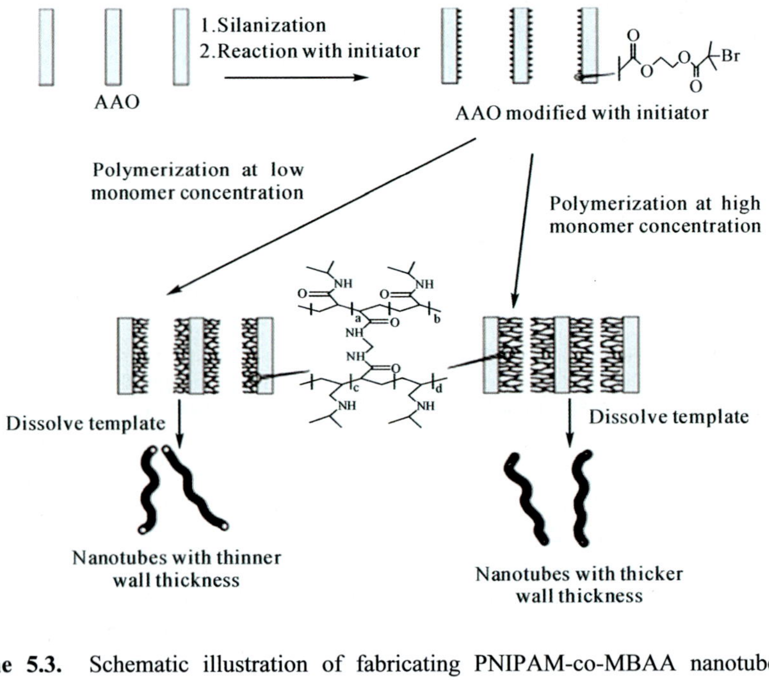

Scheme 5.3. Schematic illustration of fabricating PNIPAM-co-MBAA nanotubes with different wall thickness in a porous AAO membrane. Reprinted from Cui et al., 2006. Copyright (2006), with the permission from ACS

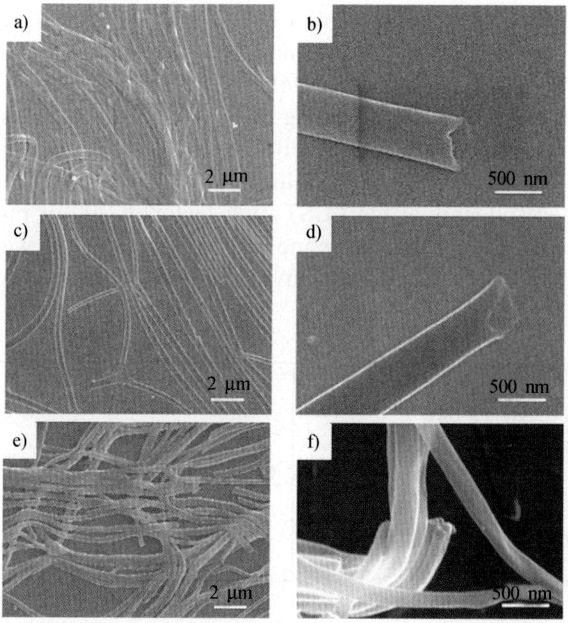

Fig.5.10. Typical SEM images of PNIPAM-co-MBAA nanotubes with different wall thicknesses. a)~c) Low-magnification of sample NT1, NT2, and NT3. d)~f) High-magnification of NT1, NT2, and NT3. Reprinted from Cui et al., 2006. Copyright (2006), with permission from ACS

Poly(*N*-isopropylacrylamide) (PNIPAM) is an interesting thermosensitive polymer (Kuckling et al., 2000). The inverse solubility-temperature relationship of this polymer in water results in the phase separation of PNIPAM solutions above the lower critical solution temperature (LCST) and thus PNIPAM is considered to be hydrophilic below the LCST or hydrophobic above the LCST. For PNIPAM in pure water, the LCST is 32 °C. PNIPAM-based polymers and derivatives are being exploited in sensors, responsive membranes, drug-delivery vehicles, and anti-fouling surfaces (Chen et al., 2001; Reber et al., 2001; Kost, 2001; Cunliffe et al., 2000). The sensitivity of the PNIPAM-co-MBAA nanotubes towards the temperature of the surrounding medium is a basic requirement for their use as thermosensitive microdevices. The temperature-induced changes in dimension and shape of the nanotubes were investigated by AFM (Jiang et al., 2003; Cui et al., 2005). Fig.5.11 shows AFM images of the exact same nanotube measured at 22, 32, and 40 °C, respectively. As the temperature increases from 22 to 40 °C, the width of the nanotube decreases from 471 to 375 nm, while the height increases from 38 to 101 nm. The nanotubes' width-to-height ratio changes show that the nanotube has an ellipse shape at 22 °C and a more circular shape at 40 °C. This should correspond to changes in the elastic modulus of the PNIPAM nanotube wall during the increase in temperature. As mentioned above, below the LCST the PNIPAM-co-MBAA nanotubes are hydrophilic and swollen, and water absorbed into the polymer acts as a plasticizer to reduce its elastic modulus dramatically (Gilcreest et al., 2004). So, at temperatures below LCST, the nanotubes collapse into a “close” shape because of gravitational effects and a low elastic modulus. When the temperature increases above the LCST, the PNIPAM-co-MBAA nanotubes shrink and become hydrophobic. At this stage, water acting as a plasticizer is extruded out of the nanotube wall so that the elastic modulus of the tube wall increases greatly. Harmon et al. (2003) proved that the elastic modulus of a low cross-linked density PNIPAM gel layer increased suddenly at the LCST by a factor of 100 or more. Hence, it is believed that the increase of the elastic modulus of the nanotube wall is responsible for the circular shape of the nanotubes above the LCST.

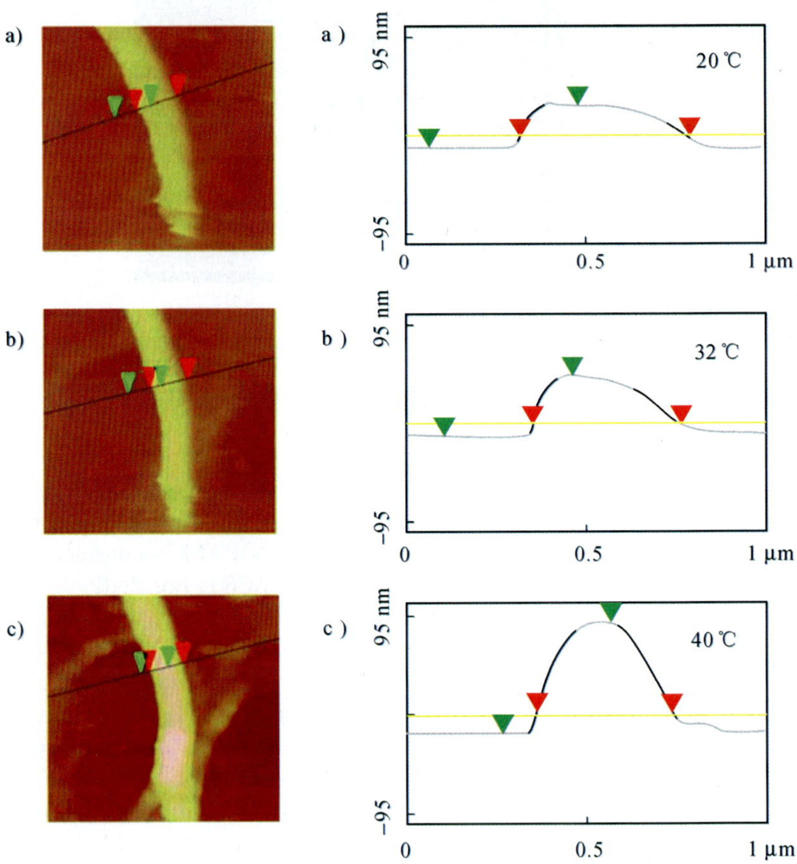

Fig.5.11. AFM images of the same nanotube in water at different temperatures: a) 22 °C; b) 32 °C; and c) 40 °C. Reprinted from Cui et al., 2005. Copyright (2005), with permission from Wiley

5.4 Functional Composite Nanotubes towards Biological Applications

Template synthesis has proven to be a versatile method to produce tubes of numerous materials and sizes. The ability to tailor the size of the nanotube for the application at hand makes nanotubes attractive for biological applications, such as drug delivery, DNA detection and gene transfection. Moreover, the shape of the tube allows for high payload capacity. With differential functionalization of the inner and outer tube surfaces, there are endless possibilities for nanotubes in biomedical nanotechnology.

5.4.1 Biofunctional and Biodegradable Nanotubes

One of the most exciting aspects of LbL technique is the fact that functional components can be introduced into these films without significant alteration of their physical and chemical properties (Li et al., 2005; Lynn, 2007; Ariga et al., 2007; Bertrand et al., 2000). Meanwhile, the LbL strategy also demonstrates that the incorporation of biomolecules in multilayer films can keep their biological activity. It is well-known that most proteins are zwitterionic macromolecules (Li et al., 2001; Willner, Katz, 2000; He et al., 2008c). Therefore, by adjusting the pH of protein solutions, one can have both positively and negatively charged proteins (below or above the isoelectric point) at different pH. In view of this, our group reported the LbL assembly of a protein human serum albumin (HSA) alone and with a phospholipid (*L*-α-dimyristoylphosphatidic acid, DMPA) in AAO templates with a diameter of about 200 nm (Lu et al., 2005). The charges of the HSA molecules for deposition were inversed by adjusting the pH of the HSA solutions used. The alternating adsorption of charged HSA (positive charge at pH 3.8 and negative charge at pH 7.0) at each deposition (neutral washing steps in between) results in the LbL deposition of a HSA multilayer in the inner walls of the AAO membrane. After the removal of the template, an array of aligned HSA nanotubes with a wall thickness of around 30 nm was obtained. Similarly, negatively charged DMPA with HSA can be a pair to assemble the phospholipids/protein nanotubes through electrostatic interaction. Alternating adsorption of HSA-DMPA layers were constructed as in the case of HSA alone (Lu et al., 2007; An et al., 2004; 2005). Circular dichroism (CD) was used to demonstrate that the preparation conditions did not destroy the overall secondary structure of HSA. However, no subsequent experiment was carried out to demonstrate the retention of biochemical activity for the protein tube in this study. For this purpose, we prepared cytochrome C (cyto-C) tubes by means of the LbL method through electrostatic attraction with PSS and a chemical reaction with glutaraldehyde (GA) on the inner wall of the PC template pores (Fig.5.12). Both the as-assembled cyto-C/GA and cyto-C/PSS nanotubes have uniform sizes and bendable shapes. The measurements of UV-visible spectra and

CD on the assembled nanotubes confirmed the cyto-C existence in the tubes (Figs.5.12b and 5.12c). Especially, the biochemical activity and electronic activity of proteins in the nanotubes were retained (Figs.5.12d and 5.12e).

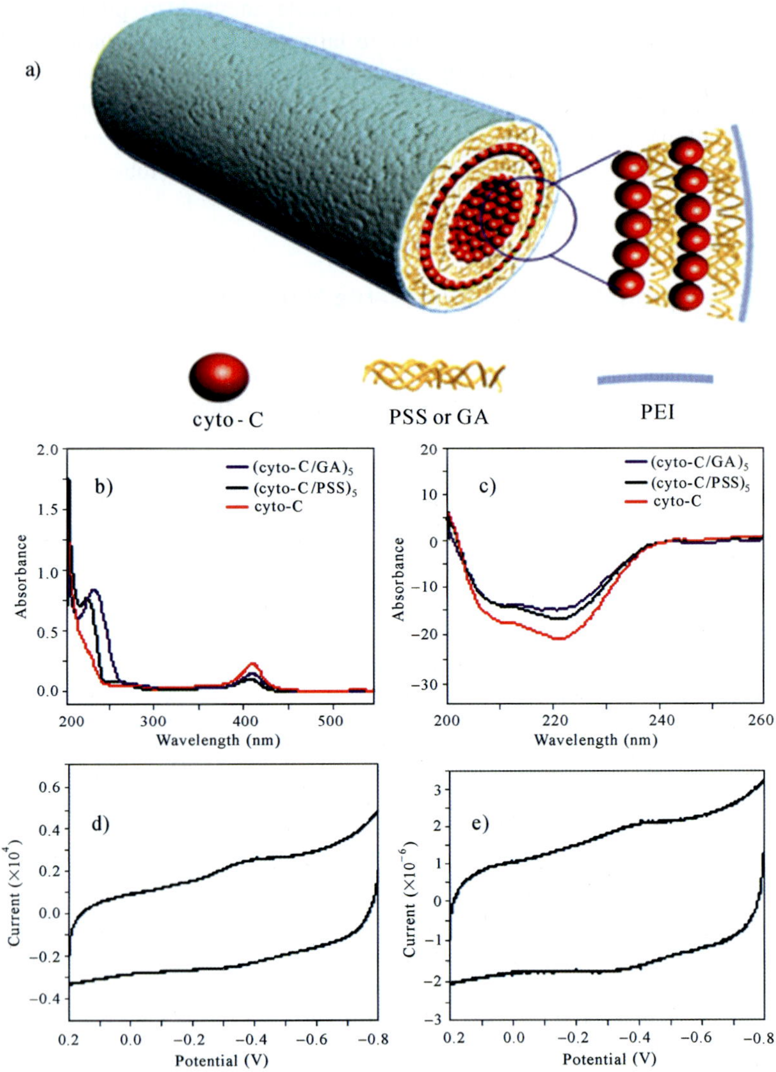

Fig.5.12. a) Schematic representation of cyto-C/GA or cyto-C/PSS nanotubes with LbL assembly on the inner wall of the alumina template based on covalent bonding or charge interaction. b) UV and c) CD spectra of (cyto-C/GA)$_5$ nanotubes, (cyto-C/PSS)$_5$ nanotubes, and pure cyto-C water solution. d) Cyclic voltammograms for the electrode after attachment of the (cyto-C/GA)$_5$ nanotubes. e) Cyclic voltammograms for the electrode after attachment of the (cyto-C/PSS)$_5$ nanotubes. Reprinted from Tian et al., 2006a. Copyright (2006), with permission from ACS

Hou et al. (2005b) reported the fabrication of glucose oxidase (GOD)/GA and hemoglobin (Hb) nanotubes using a similar strategy in the pore walls of AAO templates with a diameter of 200 nm. The activity of the GOD in the liberated nanotubes increased along with the number of protein layers in their walls. However, inside the AAO hard templates the activity of the nanotubes decreased with more than three bilayers, due to the accessibility of the protein molecules through the hollow channel inside the nanotubes became more and more limited as the diameter of the channel decreased with each additional layer. Also, Hb nanotubes that were produced in a similar manner exhibited heme electroactivity in this study.

Caruso and coworkers used porous PC membranes with pore diameters of either 400 or 100 nm as templates for the LbL deposition of peroxidase (POD)-PSS complexes [(POD-PSS)c] and oppositely charged PAH to prepare high-surface area thin films for biocatalysis (Yu et al., 2005). An average thickness of 5 nm was calculated for each (POD-PSS)c/PAH bilayer from microscopy images. The activity of the POD/PE multilayer films was found to be dependent on the amount of enzyme in the film (which is determined by the number of (POD-PSS)c layers deposited) and the total membrane surface area. Films deposited on the PC membranes with 100-nm-diameter pores showed maximum bioactivity at five (POD-PSS)c layers. Beyond this layer number, the total membrane activity decreased sharply, which is attributed to membrane pore blockage. Films deposited on PC membranes, with pore diameters of 400 nm, showed regularly increasing bioactivity up to seven (POD-PSS)c layers, with a plateau in activity observed thereafter. Activity enhancements of up to almost an order of magnitude larger were observed for the enzyme films deposited on the PC membranes (*e.g.* 100-nm pore diameter membranes with 4% porosity), compared to identical films formed on nonporous supports with the same geometrical area.

Many researchers promised that composite nanotubes have potential in the application of biological systems. However, both biodegradability and biocompatibility of these artificial nanotubes have to be taken into account. In a report by our group (Yang et al., 2007), nanotubes containing natural polyelectrolytes, alginate (ALG), and chitosan (CHI), were fabricated through electrostatic LbL assembly in porous PC membrane with a diameter of 400 nm (Fig.5.13). The assembled materials present good film formation ability and the thickness of nanotubes wall can be controlled by changing the assembled layers as one layer thickness was 5 nm. Such assembled nanotubes present good biodegradability after they are immersed in the pancreatin and low cytotoxicity because they can be internalized into cells easily (Fig.5.13b).

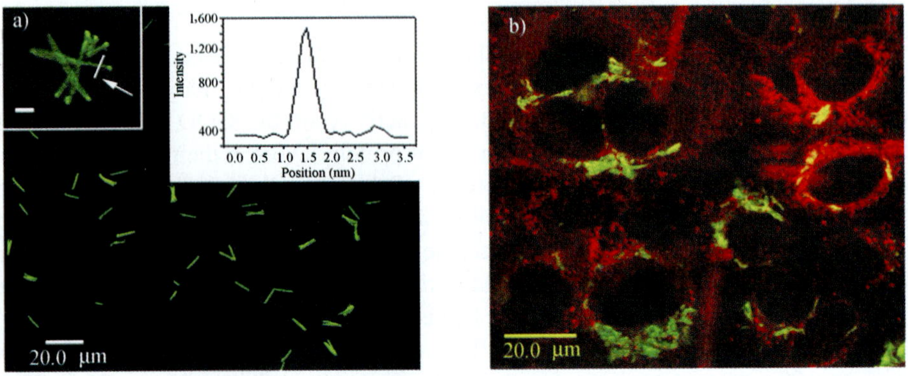

Fig.5.13. a) CLSM images of (DTAF-ALG/CHI)$_8$ nanotubes, the inset image is a selected fluorescent nanotube and its intensity profile (the inset bar is 2 μm). b) CLSM characterization of MCF-7 cells after co-culturing with (DTAF-ALG/CHI)$_8$ nanotubes for 24 h. Reprinted from Yang et al., 2007. Copyright (2007), with permission from Elsevier

5.4.2 Nanotubes for Biosensors and Bioseparation

The functionalization of nanotubes is an effective strategy towards design of new hybrid materials combining customized properties and anisotropy (Wong et al., 1997; Baughman et al., 2002). Such materials have attracted considerable interest for applications such as biocatalysts, biosensors, and as platforms for bioseparation (Williams et al., 2002; Gooding et al., 2003; Shi et al., 2006). For example, quantum dots (QD) exhibiting narrow emission bandwidth, photochemical stability, and high quantum yield have been incorporated into the walls of nanotubes (Bruchez et al., 1998). Steinhart and coworkers employed direct LbL deposition of charged dendrimers within the pores of ordered porous alumina (Kim et al., 2005). 3-aminopropyl dimethylethoxysilane (3-APDMES) was used as the first layer on the template pore wall to provide a positively charged surface, on which the first negatively charged dendrimer was directly deposited. Bilayers containing globular-shaped *N,N*-disubstituted hydrazine phosphorus-containing dendrimers of the fourth generation, having 96 terminal functional groups with either a cationic $[G_4(NH^+Et_2Cl^-)_{96}]$ or anionic $[G_4(CH\text{-}COO\text{-}Na^+)_{96}]$ character, were deposited on the walls of AAO hard templates with a diameter of 400 nm. Since dendrimers can be considered as hard spheres, nanotubes consisting of dendrimeric polyelectrolytes might be useful if swelling or deswelling needs to be minimized. The mechanical stability of the dendrimer nanotubes increased with the number of deposited bilayers. A clear dependence of the wall thickness on the number of deposition cycles was found. One bilayer thickness of the obtained nanotubes was as thin as 2 nm, which was in agreement with the film on the flat substrate. Dendrimers tubes with a high number of functional groups exposed to their environment are potential candidates for various applications, either released from or attached to the walls of a template. Later on, the same group incorporated a graded bandgap structure for thin-film configurations on smooth substrates into the walls of dendrimer nanotubes (Feng et al., 2007). The dendrimer layers acted as a rigid scaffold for the engineering of a multilayer configuration of inorganic semiconductor QD having different diameters. Taking advantage of a fluorescence resonance energy transfer (FRET) cascade from donor nanoparticles, located near the outer surface on the nanotube walls, to acceptor particles, located near the inner surface of the nanotube walls, the hybridization of DNA strands, grafted on the inner tube surface with complementary labeled DNA strands, could be detected with significantly increased sensitivity (Scheme 5.4). Furthermore, Feng et al. (2008) recently assembled different QD species in the walls of QD/dendrimer composite nanotubes by LbL deposition. Directed FRET through the graded bandgap structure resulted in significantly enhanced detection of DNA hybridization in the nanotubes combined with high selectivity.

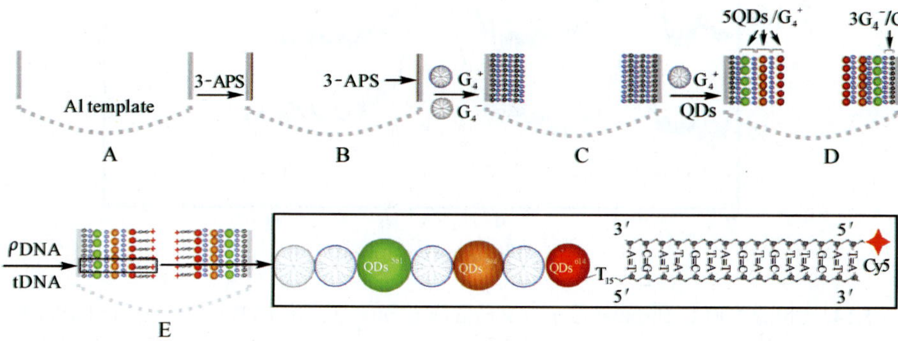

Scheme 5.4. Schematic of the nanotubes containing a graded-bandgap structure prepared by incorporating three types of ZnCdSe alloy QDs. Reprinted from Feng et al., 2007. Copyright (2007), with permission from Wiley

Differentially functionalized nanotubes can be used as smart nanophase extractors to remove specific molecules from a solution. It is a real challenge to target specific molecules to be removed from a mixture containing closely related structures. This ability was demonstrated by functionalizing the inner surface of a silica nanotube with an antibody that selects one enantiomer from a racemic pair. Enantiomers of the drug 4-[3-(4-fluorophenyl)-2-hydroxy-1-[1,2,4]triazol-1-yl-propyl]-benzonitrile (FTB) were separated from a solution by nanotubes functionalized on their inner and outer surfaces with Fab antibody fragments selective only for the RS enantiomer. This was accomplished by releasing and collecting free sol–gel prepared silica nanotubes from the template. These nanotubes were modified with an aldehyde silane, which modified both the inner and outer surfaces. Free amines on the antibody surface formed imine bonds (Lee et al., 2002). The modified nanotubes were then suspended in a solution containing a racemic mixture of FTB and stirred. Upon filtration and collection of the nanotubes, the filtrate solution was analyzed for the presence of the RS enantiomer by chiral HPLC. Results showed that 75% of the targeted enantiomer had been removed while all of the SR enantiomer remained (Figs.5.14a and 5.14b). A decrease in the racemic drug concentration from 20 to 10 μM showed that all the RS enantiomer had been selectively removed (Fig.5.14c). Unfunctionalized nanotubes did not extract a measurable amount of either enantiomer of FTB, which again demonstrates the utility of differentially functionalized nanotubes (Mitchell et al., 2002).

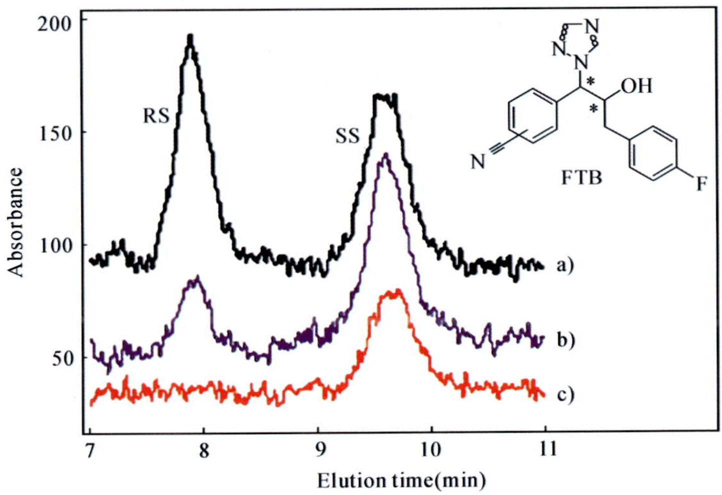

Fig.5.14. Chiral HPLC chromatograms for racemic mixtures of FTB before a), and after b), and c) extraction with 18 $mg{\cdot}mL^{-1}$ of 200-nm Fab-containing nanotubes. Reprinted from Mitchell et al., 2002. Copyright (2002), with permission from ACS

The technique of molecular imprinting allows the formation of specific recognition sites in synthetic polymers through the use of imprint molecules. These recognition sites mimic the binding sites of antibodies and other biological receptor molecules. Molecularly imprinted polymers (MIPs) can therefore be used in applications relying on specific molecular binding events. Wang and coworkers reported the synthesis of the MIP nanotube membrane using an AAO membrane by surface-initiated ATRP (Wang et al., 2006). The MIP nanotube membrane showed high affinity and selectivity in separation (Fig.5.15). Furthermore, because the molecular imprinting technique can be applied to different kinds of target molecules, ranging from small organic molecules (*e.g.* pharmaceuticals, pesticides, amino acids, nucleotide bases, steroids, and sugars) to peptides and proteins, such MIP nanotube membranes will broaden considerably the application of nanotube membranes in bioseparations and sensors.

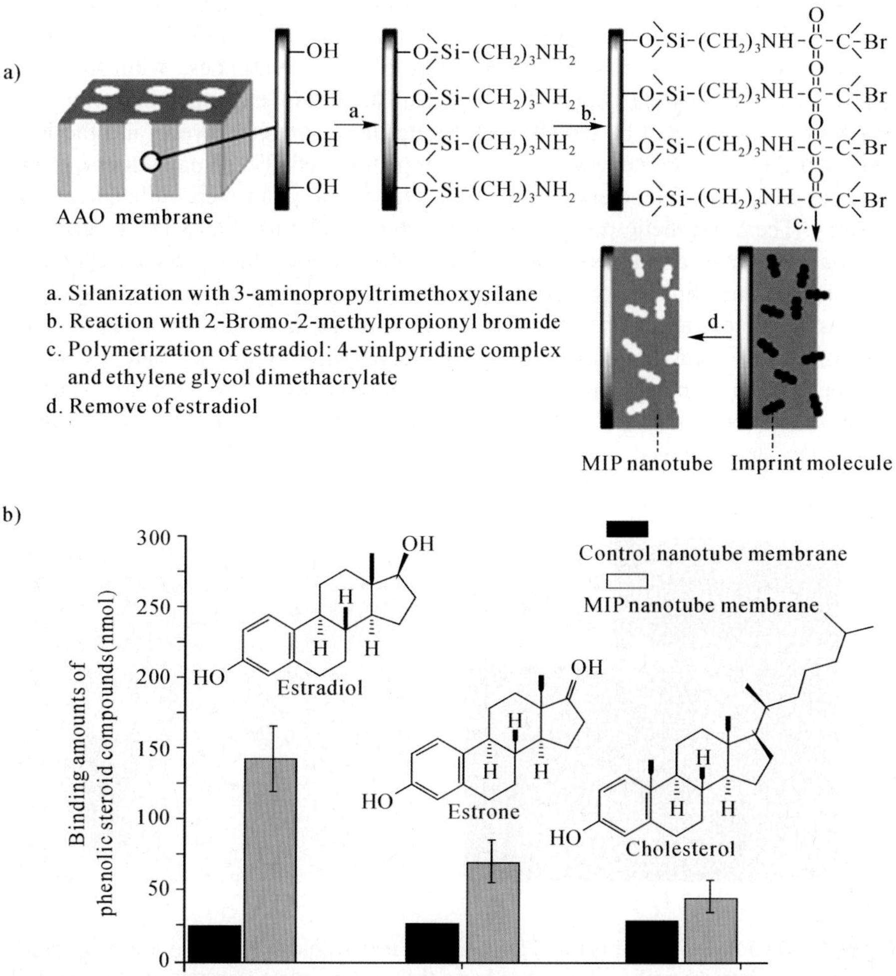

Fig.5.15. a) Scheme of template synthesized molecularly imprinted polymer nanotube membrane. b) Cross binding reactivity of the MIP nanotube membrane: a phenolic steroid compound (nmol) adsorbed by a MIP nanotube membrane from 5 mL of a 200 $\mu mol \cdot L^{-1}$ solution of phenolic steroid compound in acetonitrile in 2 h. Reprinted from Wang et al., 2006. Copyright (2006), with permission from ACS

5.4.3 Nanotubes for Drug and Gene Delivery

The ideal delivery vehicle for payloads, such as drugs, DNA, enzymes or other biomolecules, has yet to be developed. Delivery systems are designed to alleviate such issues as toxic therapeutic concentrations, failure to reach the targeted site, and poor solubility. With this in view, our group fabricated heterostructured, magnetic, polypeptide nanotubes with submicrometer dimensions by the LbL deposition of poly-*L*-lysine hydrochloride (PLL), poly-*L*-glutamic acid (PGA), and magnetic Fe_3O_4 nanoparticles within the inner pores of PC membranes, with subsequent removal of the template (Fig.5.16). The wall thickness and inner diameter of as-prepared nanotubes can be tuned by changing the assembled layers, and the length and outer diameter are dependent on the templates used. These nanotubes exhibited superparamagnetic characteristics at room temperature, and were utilized for DNA delivery. These magnetic polypeptide nanotubes can be used as a DNA carrier and manipulated under a magnetic force (Fig.5.16b). Because these tubes can be easily manipulated and directed to specific locations using magnetic forces, they can be used as separators and also carriers of drugs and agents to targeted sites *in vivo* as well as *in vitro*. That demonstrated that the combination of the template method and LbL deposition of polyelectrolytes and nanoparticles provides a versatile means to create functional nanotubes with a heterostructure that can be used for separation as well as targeted delivery.

a)

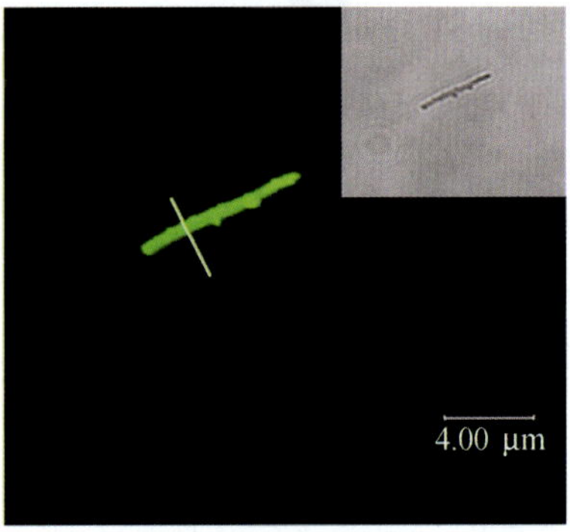

b)

Fig.5.16. a) CLSM image of plasmid DNA-conjugated magnetic nanotubes and corresponding fluorescence intensity profile. The inset shows the corresponding transmission mode. b) Photograph showing the controlled delivery of DNA-conjugated nanotubes under an external magnetic field. Reprinted from He et al., 2008b. Copyright (2008), with permission from RSC

Chen and colleagues (2005) used silica nanotubes as gene carriers. For these experiments, the inner surface of silica nanotubes were modified with 3-(aminopropyl) trimethoxysilane (APTMS) to put an overall positive charge on the membrane surface and increase loading of negatively charged DNA by electrostatic interactions. Plasmid DNA carrying the green fluorescent protein (GFP) gene was labeled with SYTO-11 to monitor DNA localization. To determine whether the nanotubes protected DNA bound to their interiors, fluorescent DNA/nanotube complex and free fluorescent DNA were treated with DNAse in separate experiments. Fluorescence intensity decreased much faster for free DNA compared with nanotube-protected DNA. The GFP plasmid loaded nanotubes were then incubated with COS-7 cells under conditions favoring nanotube uptake. Cytoplasmic

GFP expression was observed in approximately 10%~20% of the cells (Fig.5.17). Control experiments in which either nanotubes lacking plasmid DNA or free plasmid DNA was incubated with cells showed no GFP expression.

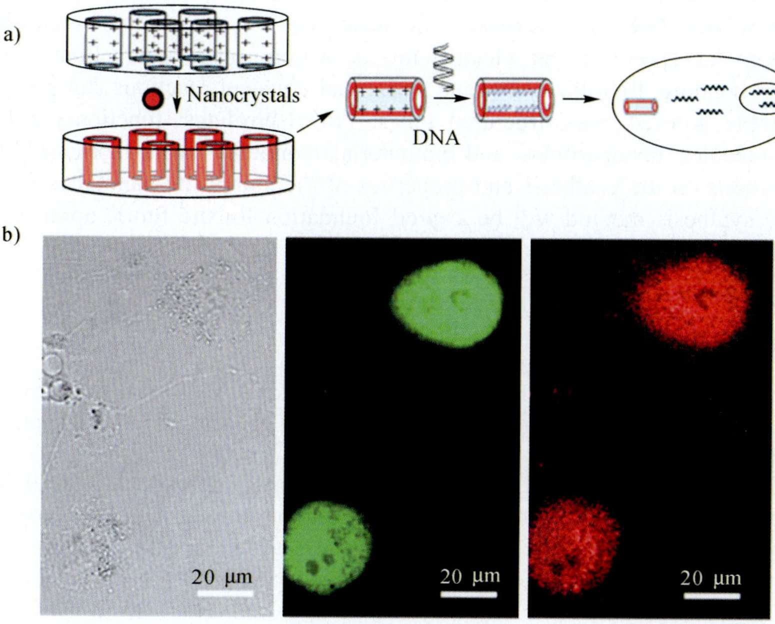

Fig.5.17. a) Schematic illustration of fluorescent nanotubes preparation and its application in gene delivery. b) Expression of GFP in cells incubated with the DNA/nanotube complex. Reprinted from Chen et al., 2005. Copyright (2005), with permission from Wiley

These studies have demonstrated the utility of template synthesized nanotubes as gene delivery vehicles. Although transfection efficiency is not high in comparison to conventional methods, such as calcium phosphate transfection, the work by Chen and colleagues established that nanotubes can be loaded with a biological agent for cellular delivery. It was made possible by the differential functionalization allowed by the template synthesis method, and it opened many avenues for using nanotubes as delivery vehicles (Hillebrenner et al., 2006).

5.5 Summary

The "porous template synthesis method" offers a simple and versatile process for the fabrication of nanotubes from various advanced functional materials. Lots of progress have been made towards the understanding of the theoretical principles underlying this method as well as towards the development of successful applications. In this review, we have focused on the fabrication of composite nanotubes using template methods combining LbL assembly, sol–gel chemistry, and polymerization, which are mainly related to biological applications. When films were fabricated within nanoporous membranes, various nanotubes can be obtained after the removal of the templates. The wall thickness of the nanotubes was proven to be controlled at the nanometer level. Except electrostatic attraction, other interactions such as hydrogen bonding, hybridizations, and sequential covalent reactions can be applied to assemble tailored, nanostructured tubes. Polyelectrolytes, functional polymer, small molecules, nanoparticles, and biomaterials nanotubes were fabricated. These investigations on the synthesis and properties of the composite nanotubes from the template synthesis method will be a good foundation for the future application of nanomaterials.

References

Aegerter MA, Mehrota RC, Oehme I, Reisfeld R, Sakka S, Wolfbeis O, Jorgensen CK (1996) Optical and electronic phenomena in sol–gel glasses and modern applications. Berlin: Springer-Verlag: 85

Ai S, He Q, Tao C, Zheng S, Li J (2005) Conductive polypyrrole and poly(allylamine hydrochloride) nanotubes fabricated with layer-by-layer assembly. Macromol Rapid Comm 20:1965-1969

Ai S, Lu G, He Q, Li J (2003) Highly flexible polyelectrolyte nanotubes. J Am Chem Soc 125:11140-11141

An Z, Lu G, Möhwald H, Li J (2004) Self-assembly of human serum albumin (HSA) and *L-α* dimyristoylphosphatidic acid (DMPA) microcapsules for controlled drug release. Chem Eur J 10:5848-5852

An Z, Tao C, Lu G, Möhwald H, Zheng S, Cui Y, Li J (2005) Fabrication and characterization of human serum albumin and *L-α*-dimyristoylphosphatidic acid microcapsules based on template technique. Chem Mater 17:2514-2519

Ariga K, Nakanishi T, Hill JP (2007) Self-assembled microstructures of functional molecules. Curr Opin Colloid Interface 12:106-120

Baughman RH, Zakhidov AA, Heer WA (2002) Carbon nanotubes: the route toward applications. Science 297:787-792

Bertrand P, Jonas A, Laschewsky A, Legras R (2000) Ultrathin polymer coatings by complexation of polyelectrolytes at interfaces: suitable materials, structure and properties. Macromol Rapid Commun 21:319-348

Bischoff BL, Anderson MA (1995) Peptization process in the sol–gel preparation of porous anatase (TiO2). Chem Mater 7:1772-1778

Bruchez M, Moronne M, Gin P, Weiss S, Alivisatos AP (1998) Semiconductor nanocrystals as fluorescent biological labels. Science 281:2013

Brumlik CJ, Martin CR (1991) Template synthesis of metal microtubules. J Am Chem Soc 113:3174-3175

Brumlik CJ, Menon VP, Martin CR (1994) Template synthesis of metal microtubule ensembles utilizing chemical, electrochemical, and vacuum deposition techniques. J Mater Res 9:1174-1183

Cai Z, Lei J, Liang W, Martin CR (1991) Molecular and supermolecular origins of enhanced electronic conductivity in template-synthesized polyheterocyclic fibrils. 1. Supermolecular effects. Chem Mater 3:960-967

Cai Z, Martin CR (1989) Electronically conductive polymer fibers with mesoscopic diameters show enhanced electronic conductivities. J Am Chem Soc 111:4138-4139

Caruso F (2000) Hollow capsule processing through colloidal templating and self-assembly. Chem Eur J 6:413-419

Chen C, Liu Y, Wu C, Yeh C, Su M, Wu Y (2005) Preparation of fluorescent silica nanotubes and their application in gene delivery. Adv Mater 17:404-407

Chen J, Yoshida M, Maekawa Y, Tsubokawa N (2001) Temperature-switchable vapor sensor materials based on *N*-isopropylacrylamide and calcium chloride. Polymer 42:9361-9365

Cui Y, Tao C, Tian Y, He Q, Li J (2006) Synthesis of PNIPAM-co-MBAA copolymer nanotubes with composite control. Langmuir 22:8205-8208

Cui Y, Tao C, Zheng S, He Q, Ai S, Li J (2005) Synthesis of thermosensitive PNIPAM-co-MBAA nanotubes by atom transfer radical polymerization within a porous membrane. Macromol Rapid Commun 26:1552-1556

Cunliffe D, Smart CA, Tsibouklis J, Young S, Alexander C, Vulfson EN (2000) Bacterial adsorption to thermoresponsive polymer surfaces. Biotechnol Lett 22:141-145

Decher G (1997) Fuzzy nanoassemblies: oward layered polymeric multicomposites. Science 277:1232-1237

Djuristic AB, Fritz T, Leo K (2000) Modeling the optical constants of organic thin films: application to 3,4,9,10-perylenetetracarboxylic dianhydride (PTCDA). Opt Commun 183:123-132

Donath E, Sukhorukov GB, Caruso F, Davis SA, Möhwald H (1998) Novel hollow polymer shells by colloid-templated assembly of polyelectrolytes. Angew Chem Int Ed 37:2202-2205

Duan L, He Q, Wang K, Yan X, Cui Y, Möhwald H, Li J (2007a) Adenosine triphosphate biosynthesis catalyzed by FoF1 atp synthase assembled in polymer microcapsules. Angew Chem Int Ed 46:6996-7000

Duan L, He Q, Yan X, Cui Y, Wang K, Li J (2007b) Hemoglobin protein hollow shells fabricated through covalent layer-by-layer technique. Biochem Biophys Res Commun 354:357-362

Feng C, Zhong X, Steinhart M, Caminade AM, Majoral J, Knoll W (2007) Graded-bandgap quantum-dot-modified nanotubes: a sensitive biosensor for enhanced detection of DNA hybridization. Adv Mater 19:1933-1936

Feng C, Zhong X, Steinhart M, Caminade AM, Majoral J, Knoll W (2008) Functional quantum-dot/dendrimer nanotubes for sensitive detection of DNA hybridization. Small 4:566-571

Ford WE, Kamat PV (1987) Photochemistry of 3,4,9,10-perylenetetracarboxylic dianhydride dyes .3. Singlet and triplet excited-state properties of the bis(2,5-di-tert-butylphenyl)imide derivative. J Phys Chem 91:6373-6380

Fu Q, Rao GVR, Basame SB, Keller DJ, Artyushkova K, Fulghum JE, Lopez GP (2004) Reversible control of free energy and topography of nanostructured surfaces. J Am Chem Soc 126:8904-8905

Gao Y, Banin U (1999) Synthesis and characterization of InAs/InP and InAs/CdSe core/shell nanocrystals. Angew Chem Int Ed 38:3692-3694

Gilcreest VP, Carroll WM, Rochev YA, Blute I, Dawson KA, Gorelov AV (2004) Thermoresponsive poly(*N*-isopropylacrylamide) copolymers: contact angles and surface energies of polymer films. Langmuir 20:10138-10145

Gooding JJ, Wibowo R, Liu J, Yang W, Losic D, Orbons S, Mearns F, Shapter JG, Hibbert DB (2003) Protein electrochemistry using aligned carbon nanotube arrays. J Am Chem Soc 125:9006-9007

Grawford GP, Steele LM, Ondris-Crawford R, Iannacchione GS, Yeager CJ, Doane JW, Finotello D (1992) Characterization of the cylindrical cavities of anopore and nuclepore membranes. J Chem Phys 96:7788-7796

Hammond PT (2004) Form and function in multilayer assembly: new applications at the nanoscale. Adv Mater 16:1271-1293

Harmon ME, Kuckling D, Frank CW (2003) Photo-cross-linkable PNIPAAm copolymers. 4. Effects of copolymerization and cross-linking on the volume-phase transition in constrained hydrogel layers. Langmuir 19:10947-10956

He Q, Cui Y, Ai S, Tian Y, Li J (2009) Self-assembly composite nanotubes and their applications. Curr Opin Colliod In 14:115-125

He Q, Song W, Möhwald H, Li J (2008a) Hydrothermal-induced structure transformation of polyelectrolyte multilayers: from nanotubes to capsules. Langmuir 24: 5508-5513

He Q, Tian Y, Cui Y, Möhwald H, Li J (2008b) Layer-by-layer assembly of magnetic polypeptide nanotubes as a DNA carrier. J Mater Chem 18:748-754

He Q, Zhang Y, Lu G, Miller R, Möhwald H, Li J (2008c) Dynamic adsorption and characterization of phospholipid and mixed phospholipid/protein layers at liquid/liquid interfaces. Adv Colloid Interface Sci 140:67-76

Hench LL, West JK (1990) The sol–gel process. Chem Rev 90:33-72

Hillebrenner H, Buyukserin F, Stewart JD, Martin CR (2006) Template synthesized nanotubes for biomedical delivery applications. Nanomedicine 1:39-50

Hou S, Wang J, Martin CR (2005a) Template-synthesized DNA nanotubes. J Am Chem Soc 127:8586-8587

Hou S, Wang J, Martin CR (2005b) Template-synthesized protein nanotubes. Nano Lett 5:231-234

Hulteen JC, Martin CR (1997) A general template-based method for the preparation of nanomaterials. J Mater Chem 7:1075-1087

Ichinose I, Senzu H, Kunitake T (1997) A surface sol–gel process of TiO_2 and other metal oxide films with molecular precision. Chem Mater 9:1296-1298

Iijima S (1991) Helical microtubules of graphitic carbon. Nature 354:56-58

Jiang Y, Yan D, Gao X, Han C, Jin X, Li L (2003) Lamellar branching of poly(bisphenol A-co-decane) spherulites at different temperatures studied by high-temperature AFM. Macromolecules 36:3652-3655

Jirage KB, Hulteen JC, Martin CR (1997) Nanotubule-based molecular-filtration membranes. Science 278:655-658

Johnston APR, Mitomo H, Read ES, Caruso F (2006) Compositional and structural engineering of DNA multilayer films. Langmuir 22:3251-3258

Johnston APR, Read ES, Caruso F (2005) DNA multilayer films on planar and colloidal supports: sequential assembly of like-charged polyelectrolytes. Nano Lett 5:953-956

Kim DH, Karan P, Göring P, Leclaire J, Caminade AM, Majoral JP, Gösele U, Steinhart M, Knoll W (2005) Formation of dendrimer nanotubes by layer-by-layer deposition.

Small 1:99-102

Kohli P, Blanchard GJ (2000) Applying polymer chemistry to interfaces: layer-by-layer and spontaneous growth of covalently bound multilayers. Langmuir 16:4655-4661

Kong H, Li W, Gao C, Yan D, Jin Y, Walton DRM, Kroto HW (2004) Poly(*N*-isopropylacrylamide)-coated carbon nanotubes: temperature-sensitive molecular nano-hybrids in water. Macromolecules 37:6683-6686

Kost J, Langer R (2001) Responsive polymeric delivery systems. Adv Drug Delivery Res 46:125-148

Kovtyukhova NI, Buzaneva EV, Waraksa CC, Martin B, Mallouk TE (2000) Surface sol–gel synthesis of ultrathin semiconductor films. Chem Mater 12:383-389

Kovtyukhova NI, Mallouk TE, Mayer TS (2003) Template surface sol–gel synthesis of SiO_2 nanotubes and SiO_2-insulated metal nanowires. Adv Mater 15:780-785

Kuckling D, Adler HJP, Arndt KF, Ling L, Habicher WD (2000) Temperature and pH dependent solubility of novel poly(*N*-isopropylacrylamide) copolymers. Marcomol Chem Phys 201:273-280

Lakshmi BB, Dorhout PK, Martin CR (1997a) Sol–gel template synthesis of semi-conductor nanostructures. Chem Mater 9:857-862

Lakshmi BB, Patrissi CJ, Martin CR (1997b) Sol–gel template synthesis of semi-conductor oxide micro- and nanostructures. Chem Mater 9:2544-2550

Lee AB, Koepsel RR, Morley SW, Matyjaszewski K, Sun Y, Russell AJ (2004) Permanent, nonleaching antibacterial surfaces. 1. Synthesis by atom transfer radical polymerization. Biomacromolecules 5:877-882

Lee DY, Nolte AJ, Kunz AL, Rubner MF, Cohen RE (2006) pH-Induced hysteretic gating of track-etched polycarbonate membranes: swelling/deswelling behavior of polyelectrolyte multilayers in confined geometry. J Am Chem Soc 128:8521-8529

Lee SB, Martin CR (2001) Controlling the transport properties of gold nanotubule membranes using chemisorbed thiols. Chem Mater 13:3236-3244

Lee SB, Mitchell DT, Trofin L, Nevanen TK, Soederlund H, Martin CR (2002) Antibody-based bio-nanotube membranes for enantiomeric drug separations. Science 296:2198-2200

Lei J, Menon VP, Martin CR (1992) Chemical preparation of conductive polypyrrole-polytetrafluoroethene composites. Polym Adv Technol 4:124-132

Li J, Möhwald H An Z, Lu G (2005) Molecular assembly of biomimetic microcapsules. Soft Matter 1:259-264

Li J, Zhang Y, Yan L (2001) Multilayer formation on a curved drop surface. Angew Chem Int Ed 40:891-894

Liang W, Martin CR (1990) Template-synthesized polyacetylene fibrils show enhanced supermolecular order. J Am Chem Soc 112:9666-9668

Liang ZJ, Susha AS, Yu AM, Caruso F (2003) Nanotubes prepared by layer-by-layer coating of porous membrane template. Adv Mater 15:1849-1853

Livage J, Henry M, Sanchez C (1988) Sol–gel chemistry of transition metal oxides. Prog Solid State Chem 18:259-341

Lu G, Ai S, Li J (2005) Layer-by-layer assembly of human serum albumin and phospholipid nanotubes based on a template. Langmuir 21:1679-1682

Lu G, Komatsu T, Tsuchida E (2007) Artificial hemoprotein nanotubes. Chem Commun 28:2980-2982

Lynn DM (2007) Peeling back the layers: controlled erosion and triggered disassembly of multilayered polyelectrolyte thin films. Adv Mater 19:4118-4130

Martin CR (1991) Template synthesis of polymeric and metal microtubules. Adv Mater 3:457-459

Martin CR (1994) Nanomaterials: a membrane-based synthetic approach. Science 266: 1961-1966

Martin CR, Parthasarathy R, Menon V (1993) Template synthesis of electronically conductive polymers: a new route for achieving higher electronic conductivities. Synth Met 55:1165-1170

Mitchell DT, Lee SB, Trofin L, Li N, Nevanen TK, Söderlund H, Martin CR (2002) Smart nanotubes for bioseparations and biocatalysis. J Am Chem Soc 124:11864-11865

Nishizawa M, Menon VP, Martin CR (1995) Metal nanotubule membranes with electrochemically switchable ion-transport selectivity. Science 268:700-702

Parthasarathy R, Martin CR (1994a) Synthesis of polymeric microcapsule arrays and their use for enzyme immobilization. Nature 369:298-301

Parthasarathy R, Martin CR (1994b) Template-synthesized polyaniline microtubules. Chem Mater 6:1627-1632

Patten TE, Maytjaszewski K (1998) Atom transfer radical polymerization and the synthesis of polymeric materials. Adv Mater 12:901-915

Peyratout CS, Daehne L (2004) Tailor-made polyelectrolyte microcapsules: from multilayers to smart containers. Angew Chem Int Ed 43:3762-3783

Quinn JF, Caruso F (2004) Facile tailoring of film morphology and release properties using layer-by-layer assembly of thermoresponsive materials. Langmuir 20:20-22

Quinn JF, Johnston APR, Such GK, Zelikin AN, Caruso F (2007) Next generation, sequentially assembled ultrathin films: beyond electrostatics. Chem Soc Rev 36:707-718

Reber N, Kuchel A, Spohr R, Wolf A, Yoshida M (2001) Transport properties of thermo-responsive ion track membranes. J Membr Sci 193:49-58

Serizawa T, Nanameki K, Yamamoto K, Akashi M (2002) Thermoresponsive ultrathin hydrogels prepared by sequential chemical reactions. Macromolecules 35:2184-2189

Shi D, Lian J, Wang W, Liu G, He P, Dong Z, Wang L, Ewing RC (2006) Luminescent carbon nanotubes by surface functionalization. Adv Mater 18:189-193

Steinhart M (2008) Supramolecular organization of polymeric materials in nanoporous hard templates. Adv Polym Sci 220:123-187

Steinhart M, Wehrspohn RB, Gösele U, Wendorff JH (2004) Nanotubes by template wetting: a modular assembly system. Angew Chem Int Ed 43: 1334-1344

Steinhart M, Wendorff JH, Greiner A, Wehrspohn RB, Nielsch K, Schilling J, Choi J, Gösele U (2002) Polymer nanotubes by wetting of ordered porous templates. Science 296:1997

Stockto WB, Rubner MF (1997) Molecular-level processing of conjugated polymers. 4. Layer-by-layer manipulation of polyaniline via hydrogen-bonding interactions. Macromolecules 30: 2717-2725

Such GK, Quinn JF, Quinn A, Tjipto E, Caruso F (2006) Assembly of ultrathin polymer multilayer films by click chemistry. J Am Chem Soc 128:9318-9319

Sukhishvili SA, Granick S (2000) Layered, erasable, ultrathin polymer films. J Am Chem Soc 122:9550-9551

Sukhishvili SA, Granick S (2002) Layered, erasable polymer multilayers formed by hydrogen-bonded sequential self-assembly. Macromolecules 35:301-310

Sun T, Wang G, Feng L, Liu B, Ma Y, Liang L, Zhu D (2004) Reversible switching between superhydrophilicity and superhydrophobicity. Angew Chem Int Ed 43:357-360

Tian Y, He Q, Cui Y, Li J (2006a) Fabrication of protein nanotubes based on layer-by-layer assembly. Biomacromolecules 7:2539-2542

Tian Y, He Q, Cui Y, Tao C, Li J (2006b) Assembly of nanotubes of poly(4-vinylpyridine) and poly(acrylic acid) through hydrogen bonding. Chem Eur J 12:4808-4812

Tian Y, He Q, Li J (2006c) Fabrication of fluorescent nanotubes based on layer-by-layer assembly via covalent bond. Langmuir 22:360-362

Tian Y, He Q, Tao C, Cui Y, Ai S, Li J (2006d) Fabrication of polyethyleneimine and poly(styrene-alt-maleic anhydride) nanotubes through covalent bond. J Nanosci Nanotechnol 6:2072-2076

Van Dyke LS, Martin CR (1990) Electrochemical investigations of electronically conductive polymers. 4. Controlling the supermolecular structure allows charge transport rates to be enhanced. Langmuir 6:1118-1123

Wang H, Zhou W, Yin X, Zhuang Z, Yang H, Wang X (2006) Template synthesized molecularly imprinted polymer nanotube membranes for chemical separations. J Am Chem Soc 128: 15954-15955

Wang K, He Q, Cui Y, Yan X, Qi W, Li J (2007) Encapsulated photosensitive drugs by biodegradable microcapsules to incapacitate cancer cells. J Mater Chem 17:4018-4021

Wang L, Wang Z, Zhang X, Shen J, Chi L, Fuchs H (1997) A new approach for the fabrication of an alternating multilayer film of poly(4-vinylpyridine) and poly(acry1ic acid) based on hydrogen bonding. Macromol Rapid Commun 18:509-514

Wang Y, Angelatos AS, Caruso F (2008) Template synthesis of nanostructured materials via layer-by-layer assembly. Chem Mater 20:848-858

Wang Z, Chumanov G (2003) WO3 sol–gel modified Ag nanoparticles arrays for electrochemical modulation of surface plasmon resonance. Adv Mater 15:1285-1289

Williams KA, Veenhuizen PT, Torre BG, Eritja R, Dekker C (2002) Nanotechnology: carbon nanotubes with DNA recognition. Nature 420:761

Willner I, Katz E (2000) Integration of layered redox proteins and conductive supports for bioelectronic applications. Angew Chem Int Ed 39:1180-1218

Willner I, Patolsky F, Wasserman J (2001) Photoelectrochemistry with controlled DNA-cross-linked CdS nanoparticle arrays. Angew Chem Int Ed 40:1861-1864

Wong EW, Sheehan PE, Lieber CM (1997) Nanobeam mechanics: elasticity, strength, and toughness of nanorods and nanotubes. Science 277:1971-1975

Yang S, Rubner MF (2002) Micropatterning of polymer thin films with pH-sensitive and cross-linkable hydrogen-bonded polyelectrolyte multilayers. J Am Chem Soc124: 2100-2101

Yang Y, He Q, Duan L, Cui Y, Li J (2007) Assembled alginate/chitosan nanotubes for biological application. Biomaterials 28: 3083-3090

Yu A, Liang Z, Caruso F (2005) Enzyme multilayer-modified porous membranes as biocatalysts. Chem Mater 17:171-175

Zelenski CM, Dorhout PK (1998) Template synthesis of near-monodisperse microscale

nanofibers and nanotubules of MoS_2. J Am Chem Soc 120:734-742
Zelikin AN, Li Q, Caruso F (2006) Degradable polyelectrolyte capsules filled with oligonucleotide sequences. Angew Chem Int Ed 45:7743-7745
Zheng S, Tao C, He Q, Zhu H, Li J (2004) Self-assembly and characterization of polypyrrole and polyallylamine multilayer films and hollow shells. Chem Mater 16: 3677-3681
Zhi L, Wu J, Li J, Stepputat M, Kolb U, Muellen K (2005) Diels-alder reactions of tetraphenylcyclopentadienones in nanochannels: fabrication of nanotubes from hyperbranched polyphenylenes. Adv Mater 17:1492-1496
Zhou Y, Shimizu T (2008) Lipid nanotubes: a unique template to create diverse one-dimensional nanostructures. Chem Mater 20:625-633

Index